ECO-FRIENDLY FARMING: HARNESSING THE POTENTIAL OF MAGGOTS

Oko Obasi

Eco-friendly Farming© Copyright 2024 Oko Obasi

ALL RIGHTS RESERVED

No part of this book may be reproduced ,copied ,stored in a retrieval system or transmitted in any form or by any means used in any manner without the prior written permission of the copyright owner, or as expressly permitted by law or under terms agreed with the appropriate reprographic rights organization. However the use of short quotation for personal or group study is permited and encouraged. Permission will also be granted upon request for the use of brief quotations in a book review.

PUBLISHED BY

Obasi Reloaded Publishing Inc.

Tel: +2348065712051

Email:obasiokoreloaded@gmail.com

ISBN: 9798329386844

Imprint: Independently published

First Edition: 2024

Printed in the United States of America.

Cover design by Oko Obasi

TABLE OF CONTENTS

Preface

Dedication

Acknowledgement

1. Introduction

   - The journey of discovering maggot farming during the COVID-19 lockdown

   - Overview of the book's purpose and structure

   - Importance of sustainable farming practices

2. Chapter 1: The Basics of Maggot Farming

   - What is maggot farming?

   - History and evolution of maggot farming

   - Types of insects used, with a focus on Black Soldier Flies

3. Chapter 2: The Science Behind Maggot Farming

   - Nutritional profile of maggots

   - The life cycle of Black Soldier Flies

   - Conversion rates and efficiency compared to traditional livestock

4. Chapter 3: Setting Up Your Maggot Farm

   - Choosing the right location

   - Necessary equipment and infrastructure

   - Sourcing and handling of initial stock

5. Chapter 4: Breeding and Growing Maggots
   - The breeding process
   - Optimizing environmental conditions (temperature, humidity, etc.)
   - Feeding practices using organic waste
6. Chapter 5: Harvesting and Processing
   - Harvesting techniques
   - Drying and processing maggots into meal
   - Storage and packaging
7. Chapter 6: Applications of Maggot Meal
   - Uses in animal feed
   - Potential for human consumption
   - Other industrial applications
8. Chapter 7: Economic and Environmental Benefits
   - Cost analysis and profitability
   - Environmental impact and sustainability
   - Case studies of successful maggot farms
9. Chapter 8: Challenges and Solutions
   - Common issues in maggot farming and how to address them
   - Regulatory considerations
   - Market challenges and opportunities
10. Chapter 9: Scaling Your Maggot Farming Business
    - Expansion strategies
    - Marketing and sales
    - Networking and industry collaboration
11. Chapter 10: Future of Maggot Farming

- Innovations and technological advancements

- Predictions and trends for the future

- The role of maggot farming in global food security

12. Conclusion

    - Recap of key points

    - Encouragement for readers to start their own maggot farming journey

    - Resources for further learning and support

13. Appendices

    - Glossary of terms

    - Resource lists (books, websites, organizations)

    - Templates and checklists for farmers

Additional Suggestions

- Interactive Elements: Incorporate QR codes or links to video tutorials and online resources.

- Training Course: Develop an online course with modules that align with each chapter of the book, offering deeper dives and practical demonstrations.

- Community Building: Create an online forum or social media group for readers and course participants to share experiences and ask questions.

DEDICATION

This book is dedicated to all the innovators, entrepreneurs, researchers, and enthusiasts who envision a sustainable future through insect farming. Your dedication to exploring new possibilities and embracing unconventional solutions inspires us all to strive for a healthier, more resilient planet.

Acknowledgments

I am deeply grateful for the immense contributions and unwavering support of several individuals who have played pivotal roles in the realization of this book and our venture into BSF maggot farming.

First and foremost, my heartfelt appreciation goes to my lovely wife, Mrs. Ifeyimwa Peace Obasi, whose dedication, vision, and passion for sustainable farming have been the driving force behind this journey. Her unwavering support and encouragement have been invaluable.

To our wonderful children—Divine, David, Richard, Victor, and Angel Michael—thank you for your patience, understanding, and unwavering belief in our goals. Your enthusiasm and love have been a constant source of inspiration.

I extend my deepest gratitude to my beloved mother, Mrs. Esther Obasi Igbokwe, whose wisdom, guidance, and unwavering support have been a source of strength throughout this endeavor.

Special recognition goes to Dr. Jane Ifeoma Okafor, the Clerk of the Senate Committee on Ecology and Climate Change. Your immense support through necessary financial and material assistance has been instrumental in bringing this book to fruition. Your belief in our vision and your contributions have made a significant impact.

I am also grateful to my mentor and leader, Hon. Nnanna Uzor Kalu, whose guidance, insight, and encouragement have been invaluable on this journey.

Lastly, to all our friends, family members, and supporters who have contributed in various ways—though too numerous to mention individually—your support has been deeply appreciated. Together, you have all played a crucial role in the success of our venture.

Thank you all for your unwavering support, belief, and contributions. This book and our journey into BSF maggot farming would not have been possible without each of you.

PREFACE

Benefits of Maggot Farming

So, what makes maggot farming so appealing? Here are some of the key benefits:

- Sustainable protein source: BSF larvae are rich in protein (up to 45%), making them an attractive alternative to traditional animal feed sources like fishmeal and soybean meal.

- Environmental benefits: Maggot farming reduces waste, utilizes organic matter, and requires minimal land, water, and feed.

- Food security: BSF larvae can be used as a direct human food source, providing a sustainable solution for global protein needs.

- Low-cost production: Maggot farming is relatively inexpensive compared to traditional animal feed production methods.

BSF maggot farming isn't just about insects; it's about addressing global challenges through innovative agriculture. From reducing organic waste to providing high-protein feed alternatives, BSF farming embodies sustainability while offering a viable business opportunity.

What You'll Discover in Our Book:

�� The Beginning: The motivation, initial research, and the decision to venture into BSF maggot farming.
�� Building the Farm: Insights into setting up infrastructure, overcoming challenges, and the pivotal role of family support.
�� Scaling Up: Strategies for increasing production, optimizing

operations, and achieving profitability.

◆◆ Market Success: How she identified buyers, established market presence, and navigated the agricultural landscape.

◆◆ Impact & Legacy: The environmental benefits, community impact, and the promising future of BSF farming.

Applications of Maggot Farming

The applications of maggot farming are diverse and far-reaching:

- Animal feed: BSF larvae are used as a protein-rich feed supplement for poultry, aquaculture, and livestock.

- Human nutrition: BSF larvae are consumed in some countries as a snack or used as an ingredient in food products.

- Organic fertilizer: Maggot frass (waste) is a valuable organic fertilizer.

Challenges and Opportunities

While maggot farming presents numerous benefits, there are challenges to overcome:

- Regulatory frameworks: Lack of standardization and regulation hinders industry growth.

- Public perception: Entomophagy and maggot farming face cultural and social acceptance barriers.

- Scalability: Large-scale production requires significant investment and infrastructure development.

- Research and development: Continued R&D is necessary to improve breeding, rearing, and processing techniques.

I invite you to join me in celebrating her journey and the valuable lessons learned along the way. Whether you're passionate about sustainable agriculture, interested in entrepreneurial narratives, or simply curious about innovative farming practices, our book promises to inspire and educate.

Stay tuned as we share more insights and updates on our journey in BSF maggot farming. Together, let's champion sustainability, entrepreneurship, and positive change. Join me in this exciting journey and discover how maggot farming can make a significant impact on your life and the world around us.

One man's meat is indeed another man's poison. Maggot farming has transformed what was once considered waste into a valuable resource, offering a sustainable solution for protein production. As the industry continues to grow and mature, it is essential to address the challenges and seize the opportunities that come with this innovative approach. Embracing maggot farming and entomophagy may just be the key to unlocking a more sustainable food future for generations to come.

INTRODUCTION

When the world went into lockdown during the COVID-19 pandemic in 2020, many of us found ourselves reevaluating our lives and looking for new opportunities. For me, this period of reflection and exploration led to a discovery that would change my life: maggot farming. Initially, it was the unusual nature of the topic that piqued my curiosity, but as I delved deeper, I realized the enormous potential and benefits it offered.

In a world increasingly focused on sustainability, maggot farming stands out as a revolutionary way to transform waste into high-quality protein. 'Eco-friendly Farming: Harnessing the Potential of Maggots' demystifies this innovative practice, guiding you through every step from setup to harvest. Maggot farming has emerged as a revolutionary aspect of agriculture, offering significant benefits to livestock and crop farmers worldwide. A few decades ago, the idea of raising maggots for profit would have seemed far-fetched. However, today, maggot farming is transforming many financially struggling individuals into successful entrepreneurs. Despite their appearance, maggots represent immense economic potential. These fly larvae are easy to feed and manage, requiring minimal inputs.

Discover the incredible efficiency and environmental benefits of maggot farming, which requires minimal resources while yielding maximum results. Learn how Black Soldier Fly larvae can convert organic waste into nutrient-rich feed, providing a sustainable alternative for livestock and aquaculture industries.

This comprehensive guide covers everything you need to know, from the science behind maggot farming to practical advice on breeding, growing, and harvesting. Whether you're an aspiring farmer, an entrepreneur, or an advocate for sustainable practices, this book will show you that beneath the surface lies a treasure trove of opportunities.

Maggot farming, specifically using Black Soldier Flies (BSF), is a revolutionary approach to addressing some of the most critical challenges in modern agriculture and waste management. In a world where sustainability is no longer just an option but a necessity, maggot farming stands out as a beacon of hope. This method transforms organic waste into high-quality protein, providing an efficient and eco-friendly alternative to traditional livestock feed.

The idea of farming maggots might seem unappealing at first glance. The sight and smell can indeed be off-putting. However, beneath this initial layer of discomfort lies a practice that is not only highly sustainable but also incredibly profitable. By turning waste into a valuable resource, maggot farming offers a solution that benefits both the environment and the economy.

Don't let the sight and the smell distract you. There is a pot of gold here. Maggots are produced from waste, making their farming an excellent integration for sustainable agriculture. In essence, no waste means no maggots. Every farmer's goal is to maximize profit, and since feeding constitutes about 70% of animal production costs, reducing this expense can lead to higher profits. One effective way to achieve this is through maggot production.

Maggots, derived from domestic flies, are rich in protein and other essential nutrients that animals need for their daily activities. Setting up a maggot farm, or maggotery, can be cost-effective and requires suitable facilities for mass production. Many livestock farmers, especially those raising fish and poultry, now incorporate maggots—either dried or fresh—into their animal feed or use them directly as feed.

In "Eco-friendly Farming: Harnessing the Potential of Maggots," you will find a comprehensive guide that covers all aspects of maggot farming. This book is designed to be accessible to everyone, whether you are an experienced farmer, a newcomer to sustainable practices, or an entrepreneur looking for a new venture. Each chapter takes you through the various stages of maggot farming, from understanding the science behind it to practical steps on setting up your farm, breeding and growing maggots, and harvesting and processing them.

We will explore the remarkable efficiency of maggots in converting organic waste into protein-rich biomass. You will learn about the life cycle of the Black Soldier Fly, how to create optimal conditions for their growth, and the numerous applications of maggot meal in animal feed and beyond. Additionally, we will discuss the economic and environmental benefits, as well as the challenges and solutions you may encounter along the way.

Maggot farming is more than just a novel idea; it is a crucial part of the sustainable future we must build. By embracing this practice, we can contribute to a more efficient, eco-friendly, and profitable agricultural system. I invite you to join me on this journey and discover the hidden potential that maggot farming holds.

Let's look past the surface and see the pot of gold that lies beneath. Welcome to the future of sustainable farming.

CHAPTER 1: THE BASICS OF MAGGOT FARMING

Introduction to Maggot Farming

Maggot farming, a term that might initially conjure up images of decay and unpleasantness, is in reality a groundbreaking approach to sustainable agriculture and waste management. This practice involves the cultivation of larvae, particularly those of the Black Soldier Fly (BSF), for various purposes, including animal feed, composting, and even potential human consumption. In this chapter, we will explore the fundamental concepts of maggot farming, its historical evolution, and why it is gaining traction as a sustainable solution for the future.

What is Maggot Farming?

At its core, maggot farming is the process of breeding and rearing insect larvae to convert organic waste into protein-rich biomass. The larvae of the Black Soldier Fly are particularly well-suited for this purpose due to their rapid growth rate, high protein content, and ability to thrive on a wide variety of organic materials. These larvae can be used to produce high-quality animal feed, effectively addressing the need for sustainable protein sources in agriculture.

As the world faces growing challenges of land and water scarcity for traditional crop cultivation, Black Soldier Fly (BSF) larvae have emerged as a revolutionary solution to transform the global food system. These remarkable organisms possess the ability to convert a wide range of organic waste streams into highly valuable protein, fats, and minerals.

BSF larvae are nutritional powerhouses, boasting an impressive nutrient profile. Dried BSF larvae can contain up to 50% protein, 35% fat (with an amino acid composition comparable to fishmeal), 6% calcium, 1.2% phosphorus, 1% magnesium, and 0.3% sodium. This exceptional nutritional density has earned BSF larvae recognition and widespread use as high-quality protein and fat sources in the feeds for poultry, pigs, chickens, fish, shrimp, and prawns, serving as a sustainable alternative to conventional commercial fishmeal.

The versatility and efficiency of BSF larvae in upcycling organic waste into nutrient-rich biomass offer a promising pathway to address the growing global demand for animal feed and protein. By leveraging this natural bioconversion process, the food system can minimize its reliance on land and water-intensive crop production, while simultaneously reducing waste and environmental impact.

History and Evolution of Maggot Farming

The practice of using insects for waste management and as a food source is not new. Various cultures have utilized insects in traditional diets and agricultural systems for centuries. However, the modern concept of maggot farming, particularly with a focus on the Black Soldier Fly, has gained prominence over the last few decades. Researchers and entrepreneurs have recognized the potential of BSF larvae to address multiple global challenges, including waste management, food security, and sustainable agriculture.

Maggot farming has been used in the past and present for a variety of reasons, including

- Decaying flesh and organic matter removal

- Food source for animals

- Nutritional value comparable to beef and chicken

- Environmental benefits from energy-efficient farming practices

- Waste management

- Potential source of protein for human consumption

Modern maggot farming often uses black soldier flies and has multiple applications across the animal feed, composting, and pharmaceutical industries.

Maggot farming has gained attention globally as a sustainable and economically viable practice. Let's explore its history and its current status in Nigeria:

1. Global Perspective:

   - AgriProtein: A South African company called AgriProtein pioneered maggot farming in Africa. They established the world's largest fly farm, where they extract maggots to produce livestock feed. This initiative propelled the use of insects as a protein source beyond academic theory and into a commercial venture.

   - Environmental Benefits: Insects, including maggots, promise to be an environmentally friendly alternative to traditional livestock protein feeds. They leave a smaller ecological footprint and can provide high-quality protein for animals.

   - Research and Adoption: Studies have shown that maggot meal can serve as a potential protein source in poultry nutrition. Additionally, maggot farming has been studied extensively for fish and crustacean feed in pond farming since the late 2000s.

2. Nigeria's Opportunity:

- Untapped Potential: Despite its global success, maggot farming is relatively unexplored in Nigeria. Entrepreneurs who embrace this opportunity can potentially make millions.

- Cost-Effective Feed: Maggots can be used to feed various animals, including fish, poultry birds, ducks, pigs, and guinea fowl. By rearing maggots, farmers can significantly reduce feed costs, which is a major challenge in livestock farming.

- Low Capital Requirement: Starting a maggot farm requires minimal capital—around ₦50,000 to ₦200,000—after receiving training from an expert.

In summary, maggot farming is not only a breakthrough business but also a sustainable solution to rising animal feed costs. As awareness grows, more Nigerian farmers may recognize its potential and contribute to this innovative industry.

Why Black Soldier Fly Maggot Farming?

The Black Soldier Fly (Hermetia illucens) is native to the Americas but has spread to many parts of the world. It is known for its efficiency in converting organic waste into biomass. Unlike common houseflies, adult Black Soldier Flies do not feed and are not vectors of disease, making them an ideal choice for farming. The larvae are voracious eaters, capable of consuming large amounts of organic waste and converting it into protein and fat-rich biomass in a matter of days.

1. Sustainability:

Maggot farming offers a highly sustainable method of producing protein. The larvae can be fed on a wide range of organic waste materials, including food scraps, agricultural residues, and even manure. This not only reduces the burden on landfills but also transforms waste into a valuable resource.

2. Efficiency:

Black Soldier Fly larvae are incredibly efficient at converting nutrients. They require significantly less space, water, and feed compared to traditional livestock. Studies have shown that BSF larvae can convert organic waste into body mass at a rate four to ten times more efficient than conventional animals like cattle, pigs, or poultry.

3. High Nutritional Value:

BSF larvae are rich in protein, fats, and other essential nutrients, making them an excellent ingredient for animal feed. The larvae typically consist of 40% to 65% protein, which is comparable to or even higher than many traditional protein sources.

4. Reduced Environmental Impact:

Maggot farming has a lower environmental footprint compared to traditional livestock farming. It generates fewer greenhouse gases and requires less land and water. Additionally, the process of converting organic waste into biomass helps mitigate the impact of waste disposal.

The Black Soldier Fly: A Perfect Candidate

The Black Soldier Fly is uniquely suited for maggot farming due to its biological and ecological characteristics:

- Life Cycle: The life cycle of the BSF is short, with the larval stage lasting around two weeks under optimal conditions. This rapid growth cycle allows for quick and efficient biomass production.

- Diet: BSF larvae are detritivores, meaning they feed on decaying organic matter. They can consume a wide variety of organic waste, including food scraps, manure, and agricultural residues, making them versatile and efficient waste converters.

- Non-Pest: Adult Black Soldier Flies does not have a mouth so do not feed and do not transmit diseases, unlike common houseflies. This makes them a safe and hygienic option for farming.

Setting the Stage for Success

As we delve deeper into the world of maggot farming, it is essential to understand the basic requirements and conditions necessary for success. From setting up the farm and breeding the flies to managing the larvae and harvesting the biomass, each step requires careful planning and execution. In the subsequent chapters, we will explore these aspects in detail, providing practical guidance and insights to help you embark on your maggot farming journey.

Maggot farming is more than just an innovative idea; it is a transformative practice that has the potential to revolutionize agriculture and waste management. By harnessing the power of the Black Soldier Fly, we can create a more sustainable and efficient food production system, turning waste into wealth and ensuring a brighter future for all.

Unveiling a New Era of Prosperity

In the annals of agricultural innovation and environmental sustainability, Black Soldier Fly (BSF) farming emerges not only as a solution to waste management but as a harbinger of economic prosperity reminiscent of the transformative impact of crude oil. This chapter serves as an introduction to the promising world of BSF farming, exploring its multifaceted benefits and potential to revolutionize global agriculture.

Types of Maggots You Can Farm

While there are numerous insect larvae that can be farmed for various purposes, the most common and commercially viable maggots include:

1. Black Soldier Fly (Hermetia illucens) Larvae

   - Uses: Animal feed, composting, potential human consumption

   - Advantages: High protein content (40%-65%), rapid growth cycle, ability to consume a wide range of organic waste, low environmental impact, and non-pest status of adults.

2. Mealworm (Tenebrio molitor) Larvae

   - Uses: Animal feed, pet food, human consumption

- Advantages: High protein and fat content, suitable for human consumption, easier to manage at a small scale.

3. Common Housefly (Musca domestica) Larvae

- Uses: Traditionally used as fish bait and in some animal feeds

- Advantages: High reproduction rate, can thrive on various organic materials.

- Disadvantages: Not typically farmed on a large scale for commercial purposes due to significant drawbacks.

4. Green Bottle Fly (Lucilia sericata)

Efficient at breaking down organic matter and converting it into high-protein larvae

Commonly used in wound healing therapies (maggot debridement therapy)

5. Yellow Mealworm (Tenebrio molitor)

Not a true fly, but the larval stage of a beetle species

Considered a good alternative to traditional fly larvae in some applications

6. Indian Meal Moth (Plodia interpunctella)

Another non-fly species, the larval stage of this moth is sometimes used in maggot farming.

Can be reared on a variety of stored grain and processed food products.

7. Blue Bottle Fly (Calliphora vomitoria)

Produce larger larvae that can be used for specific applications, such as bait, pet food, or medical maggot therapy. Prefer cooler temperatures compared to other fly species, making them more suitable for certain climates.

Effectively break down organic matter and convert it into high-protein biomass.

The choice of fly species for a maggot farming operation will depend on factors such as climate, target applications, waste stream composition, and local regulations. It's important to research and select the species that best fits your specific needs and conditions.

Why We Don't Farm the Common Housefly Maggot

The common housefly maggot, while having some utility, is generally not farmed for large-scale commercial purposes due to several reasons:

1. Disease Vector: Common houseflies are known vectors of numerous diseases. They can carry and transmit pathogens like bacteria, viruses, and parasites, making them a health risk, especially when used in food production systems.

2. Hygiene Issues: Houseflies often breed in unsanitary conditions, such as decaying organic matter and feces, increasing the risk of contamination. This makes them less desirable for applications where hygiene and safety are critical.

3. Adult Fly Problems: Unlike Black Soldier Flies, adult houseflies are nuisances. They can infest human living spaces and are associated with poor sanitation. Managing the adult population can be challenging and poses a significant drawback.

4. Lower Nutritional Value: While housefly larvae can be protein-rich, they generally have a lower nutritional profile compared to Black Soldier Fly larvae, making them less ideal for high-quality animal feed.

5. Regulatory Hurdles: Due to their status as disease vectors and their association with unsanitary conditions, there are stricter regulations and restrictions on farming and using housefly larvae in many regions, adding to the complexity and cost of production.

Why Black Soldier Fly Larvae are Preferred

1. Non-Pest Adult Stage: Adult Black Soldier Flies do not feed, do not transmit diseases, and do not invade human spaces, making

them far less of a nuisance and a risk compared to houseflies.

2. High Efficiency: BSF larvae are exceptionally efficient at converting organic waste into protein-rich biomass, which is critical for creating sustainable feed solutions.

3. Versatility: They can be fed a wide range of organic waste materials, including agricultural residues, food scraps, and even manure, making them highly versatile and beneficial for waste management.

4. Environmental Impact: BSF farming has a lower environmental footprint, requiring less space, water, and resources while producing fewer greenhouse gases compared to traditional livestock farming.

5. Nutritional Profile: BSF larvae are rich in protein, fats, and other essential nutrients, making them an excellent source of feed for livestock, poultry, and aquaculture.

Conclusion

While there are various types of maggots that can be farmed, the Black Soldier Fly larvae stand out due to their numerous advantages over other options like the common housefly maggot. The non-pest nature, high efficiency in converting waste to biomass, superior nutritional profile, and lower environmental impact make BSF larvae the preferred choice for sustainable maggot farming. By understanding these differences, we can better appreciate the potential of maggot farming to revolutionize waste management and protein production in an eco-friendly manner.

Composition of BSF Maggot

The composition of Black Soldier Fly (BSF) maggots includes:

1. Protein: 42-45% (high-quality protein with all essential amino acids)

2. Fat: 25-30% (rich in unsaturated fatty acids)

3. Calcium: 3-5% (higher than milk or beef)

4. Phosphorus: 2-3% (essential for bone health)

5. Potassium: 2-3% (important for heart health)

6. Magnesium: 1-2% (involved in muscle and nerve function)

7. Sodium: 0.5-1% (lower than most animal feeds)

8. Fiber: 5-7% (prebiotic fiber supporting gut health)

9. Moisture: 5-7% (low moisture content, making them shelf-stable)

10. Ash: 2-3% (mineral content)

11. Vitamins: B1, B2, B5, B6, and B12 (important for energy metabolism and nerve function)

12. Antioxidants: Polyphenols and carotenoids (protect against oxidative stress)

The Exceptional Fatty Acid Profile of Black Soldier Fly Larvae

The fat content of Black Soldier Fly (BSF) larvae is far more qualitative than that of traditional feed fats. This distinction is largely attributed to the high concentration of lauric acid, which accounts for 53% of the total fat content. Lauric acid is a valuable substance that enhances the absorption of nutrients by the animal.

Beyond lauric acid, the fat profile of BSF larvae also contains AWME, a compound believed by Russian scientists to possess potent antibacterial properties. AWME is capable of disrupting the lipid-based structures of viruses similar to HIV and measles, as well as destroying pathogenic bacteria like Clostridium and harmful protozoa. This powerful antimicrobial activity helps to protect the intestinal health and development of animals consuming BSF larvae.

The fatty acid composition of BSF larvae is remarkably similar to that of fish oil, which is highly beneficial for the skin, coat, and brain health of aquatic species. When fish are fed BSF larvae, they exhibit accelerated growth, improved overall health, and enhanced vitality.

The comprehensive fatty acid profile of BSF larvae includes:

- Lauric Acid (53%)
- Myristic Acid
- Pentadecanoic Acid (ISO)
- Pentadecanoic Acid (Anteiso)
- Palmitic Acid
- Stearic Acid
- Palmitoleic Acid
- Oleic Acid
- Linoleic Acid

This exceptional nutritional profile, rich in beneficial fats and antimicrobial compounds, underscores the profound value of BSF larvae as a premium feed ingredient for a wide range of animal species, contributing to their overall health, growth, and well-being.

The exact composition may vary depending on factors like feed, breeding conditions, and processing methods. BSF maggots are considered a nutrient-rich food source for humans and animals.

The nutritional composition of BSF larvae will vary when **BSFLs are fed with different food sources**. So the nutrient composition of **BSF larvae** is variable. So to know the nutritional composition of the **black soldier fly more or less you have to know which food sources to eat.** The table that compares the nutritional composition of BSF varies greatly based on the food Larvae eat:

The Table of The Diverse Dietary Sources that Influence the Nutritional Profile of Black Soldier Fly Larvae:

FOOD SOURCES	% PROTEIN	% FAT
Animal Manure	39,1 % – 45,7%	29 – 35,1 %
Food Wastes	39 – 41,8 %	27,2 – 35,1 %
Green Waste	31,2 – 36,4%	5,2 – 6,63%
Raw Rice Bran	42,3 – 45,7%	27,5 – 27,8%

Black Soldier Fly (BSF) larvae exhibit an off-white coloration with distinct grayish-brown stripes, and their size can range from 0.25 to 0.5 cm in length, with a diameter of 0.1 to 0.15 cm. The remarkable versatility of these larvae lies in their ability to thrive on a wide array of organic materials, transforming them into nutrient-rich biomass.

BSF larvae are adept at upcycling a diverse range of organic waste streams, including:

- Vegetable and fruit peels
- Food processing by-products
- Animal manures
- Feathers
- Poultry or fish waste

By harnessing the larvae's exceptional bioconversion capabilities, these organic waste materials are converted into a valuable source of protein, fats, and other essential nutrients.

The specific nutrient profile of the BSF larvae is directly influenced by the composition of their dietary substrates. The nutritional value can be tailored and optimized by carefully selecting and managing the organic waste streams used to feed the larvae, allowing for the production of a highly nutritious and versatile feed ingredient.

This adaptability makes BSF larvae an increasingly valuable resource in the quest for sustainable and efficient food production systems, as they can transform a wide range of organic waste streams into high-quality, nutrient-dense biomass for use in animal feed and other applications.

The Rise of BSF Farming

BSF farming represents a paradigm shift in sustainable agriculture, leveraging nature's own efficient bioconverters to tackle pressing challenges. Similar to how crude oil revolutionized energy sectors, BSF larvae are adept at converting organic waste into nutrient-rich biomass. This capability not only addresses environmental concerns by reducing waste volumes but also provides a valuable resource for various industries, from animal feed to biodegradable materials.

Economic Potential and Wealth Creation

Drawing parallels to the economic windfall brought by crude oil, BSF farming presents new avenues for wealth creation at both individual and national levels. The larvae's high protein content and nutritional value make them a cost-effective alternative in animal feed production, potentially reducing dependence on costly imports like fishmeal. Moreover, the scalability and efficiency of BSF farming offer promising economic opportunities, from small-scale entrepreneurial ventures to large-scale industrial applications.

Sustainability at its Core

Beyond economic benefits, BSF farming champions sustainability. It requires minimal land and water compared to traditional livestock farming, thereby mitigating environmental impacts such as deforestation and water scarcity. By converting organic waste into valuable resources, BSF farming supports a circular economy model, promoting resource efficiency and reducing greenhouse gas emissions—a crucial step towards achieving global sustainability goals.

Embracing the Future

As we embark on this exploration of BSF farming, we recognize its potential to shape a more sustainable and prosperous future. By harnessing the innate abilities of BSF larvae and embracing innovative farming practices, individuals and nations can pave the way towards a resilient agricultural sector and thriving economies. This chapter sets the stage for deeper insights into BSF farming's applications, challenges, and transformative impact in subsequent chapters.

In the chapters ahead, we will delve into the practical aspects of BSF farming—from breeding and management techniques to emerging applications and global trends. Together, let us explore how BSF farming is not just a solution for today's challenges but a catalyst for sustainable development and prosperity in the decades to come.

CHAPTER 2: THE SCIENCE BEHIND MAGGOT FARMING

Understanding the science behind maggot farming is crucial for successfully harnessing its potential. This chapter delves into the biological and ecological aspects of maggot farming, focusing primarily on the Black Soldier Fly (BSF). We will explore the nutritional profile of maggots, the life cycle of the BSF, and the mechanisms that make maggot farming an efficient and sustainable practice. Maggot farming offers several benefits, including the ability to recycle organic waste, produce a high-protein animal feed ingredient, and generate revenue.

The maggots' voracious appetite and rapid growth make them a promising solution for sustainable waste management and feed production. Overall, the science of maggot farming involves understanding the biology and ecology of fly species, optimizing the rearing environment, and efficiently harvesting and processing the maggot biomass for various applications.

The Nutritional Profile of Maggots

Maggots, especially those of the Black Soldier Fly, are nutritional powerhouses. Their composition makes them an ideal source of protein and other essential nutrients for various applications, particularly in animal feed.

1. Protein Content:

   - BSF larvae contain between 40% to 65% protein by dry weight. This high protein content is comparable to traditional protein sources such as fishmeal and soymeal.

   - Proteins are crucial for growth and development in animals, making BSF larvae an excellent ingredient for feed.

2. Fat Content:

   - The larvae also have a high fat content, ranging from 15% to 35%. These fats are a source of energy and essential fatty acids necessary for animal health.

   - The lipid profile of BSF larvae includes beneficial fatty acids like lauric acid, which has antimicrobial properties.

3. Minerals and Vitamins:

   - BSF larvae are rich in essential minerals such as calcium, phosphorus, and magnesium, which are vital for bone health and metabolic functions.

   - They also contain important vitamins, including B-vitamins, which play a role in energy metabolism and overall health.

4. Fiber:

   - The chitin found in the exoskeleton of larvae acts as a source of dietary fiber. Although animals cannot digest chitin, it can contribute to gut health and aid in digestion.

5. Carbohydrates: Vital for Energy and Structure

Energy Provision: Carbohydrates serve as the primary energy source for BSF larvae, fueling essential functions such as movement, digestion, and metabolism to maintain the larvae's activity and health.

Dietary Balance: While proteins are crucial for growth, a balanced intake of carbohydrates is equally important. This balance ensures that proteins are used for growth rather than energy production, thereby reducing unnecessary ammonia emissions.

Utilization of Organic Matter: Carbohydrates present in organic waste materials, including cellulose and lignin, are valuable resources that BSF larvae efficiently break down and utilize. This capability enables the conversion of a diverse array of organic substrates into biomass.

Chitin Production: Carbohydrates form the structural basis for certain amino acids and contribute to the synthesis of chitin, a polysaccharide crucial for the formation of the larvae's exoskeleton.

Here are some interesting and fun facts about the science of Black Soldier Flies (BSF):

1. Non-feeding Adults: Unlike many other fly species, adult Black Soldier Flies do not feed. Their sole purpose as adults is to reproduce. This unique trait reduces the nuisance factor associated with adult flies and makes BSFs more hygienic.

2. Efficient Waste Converters: BSF larvae are incredibly efficient at converting organic waste into biomass. They can consume a wide variety of organic materials, including kitchen scraps, manure, and agricultural residues, and convert them into nutrient-rich larvae within a short period.

3. Rapid Growth: BSF larvae undergo rapid growth during their larval stage. Under optimal conditions, they can reach full size in about two weeks. This fast growth rate contributes to their efficiency in biomass production.

4. High Protein Content: The larvae of Black Soldier Flies are remarkably high in protein, typically ranging from 40% to 65% on a dry weight basis. This makes them a valuable source of protein

for animal feed and potentially for human consumption in the future.

5. Chitin Content: The exoskeleton of BSF larvae contains chitin, a polysaccharide that is also found in the exoskeletons of insects like crabs and shrimp. Chitin has various industrial applications, including in biomedical products and water treatment.

6. Minimal Environmental Impact: Compared to traditional livestock farming, BSF farming has a significantly lower environmental footprint. It requires less water, land, and feed, and produces fewer greenhouse gas emissions, making it a sustainable alternative for protein production.

7. Pupal Stage Dormancy: BSF pupae enter a stage of dormancy known as diapause, where their development is temporarily halted. This adaptation helps them survive adverse environmental conditions and ensures synchronized emergence of adult flies.

8. Bioconversion Efficiency: BSF larvae have an impressive bioconversion rate, meaning they efficiently convert the organic matter they consume into biomass. This efficiency contributes to their role in waste management and sustainable agriculture.

9. Attractive to Fish: BSF larvae are commonly used as feed for fish due to their high nutritional value and palatability. They provide essential nutrients to fish and help improve growth rates in aquaculture settings.

10. Research and Innovation: Ongoing research into BSFs continues to explore their potential applications, including biodegradation of plastics, extraction of valuable compounds, and even as a potential protein source for human consumption in the form of protein powders or supplements.

These fun facts highlight the fascinating biology and practical applications of Black Soldier Flies, showcasing why they are gaining attention as a sustainable solution in agriculture and waste management.

The Life Cycle of the Black Soldier Fly

Understanding the life cycle of the Black Soldier Fly is essential for effective maggot farming. The life cycle consists of several stages: egg, larva, pupa, and adult. Each stage requires specific conditions to optimize growth and productivity.

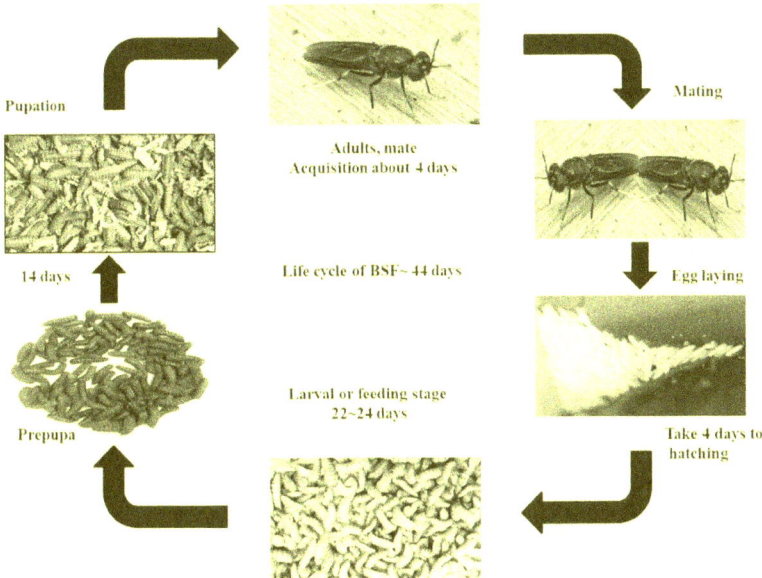

1. Egg Stage:

   - Female BSFs lay eggs in clusters, typically near decaying organic matter. Each female can lay up to 500-900 eggs in her lifetime.

   - Eggs hatch within 4-5 days under optimal conditions, releasing tiny larvae.

2. Larval Stage:

   - The larval stage is the primary period of growth and nutrient accumulation. Larvae feed voraciously on organic waste, converting it into body mass. Black soldier fly maggots go through 6-7 larval instars, during which they rapidly grow in size and weight.

The larval stage lasts 2-4 weeks, depending on factors like temperature and food availability

- This stage lasts approximately 14-18 days, during which larvae increase significantly in size and weight.

- Optimal conditions for larval growth include temperatures between 25°C and 30°C (77°F to 86°F) and high humidity levels.

3. Pupal Stage:

- Once larvae reach their maximum size, they enter the pupal stage, where they undergo metamorphosis into adult flies. When the maggots reach their final larval instar, they stop feeding and enter the pupal stage. The pupae are brown and immobile, undergoing a dramatic transformation into the adult fly.

This pupal stage lasts 2-3 weeks, during which the fly develops its wings, legs, and other adult features.

- Pupae do not feed and require a dry, secure environment to develop. This stage lasts about 7-10 days.

4. Adult Stage:

- Adult BSFs emerge from the pupae and focus on reproduction. They do not feed during this stage, relying on the energy reserves accumulated during the larval stage.

- The adult lifespan is relatively short, typically around 5-8 days. During this time, they mate and lay eggs, completing the life cycle.

Reproduce and die

1. Male BSF Adults: Male Black Soldier Flies reach adulthood primarily for the purpose of reproduction. They do not feed during this stage and typically have a short lifespan, ranging from a few days to a couple of weeks. After mating, male BSFs will die shortly thereafter.

2. Female BSF Adults: Female Black Soldier Flies also have a

relatively short adult lifespan, typically around 5-8 days. During this time, their main activity is to lay eggs. A female BSF can lay a significant number of eggs, ranging from 500 to 900, depending on conditions and her health. After completing the egg-laying process, the female BSF will also die.

3. Life Cycle Completion: The life cycle of the BSF is completed when the eggs laid by the female hatch into larvae. The larvae feed voraciously on organic matter, grow rapidly, pupate, and eventually emerge as adult flies, continuing the cycle.

This short adult lifespan and singular focus on reproduction are common traits among many fly species, including the Black Soldier Fly. It reflects their evolutionary adaptation to maximize reproductive success within a short period, often in environments rich in organic matter where their larvae can thrive.

Mechanisms of Efficiency

The efficiency of maggot farming, particularly with BSFs, is attributed to several key mechanisms:

1. Rapid Growth and High Feed Conversion:

   - BSF larvae grow rapidly, reaching full size in about two weeks. Their ability to convert organic waste into body mass is significantly higher than traditional livestock.

   - The feed conversion ratio (FCR) for BSF larvae is much lower than that of cattle, pigs, and poultry, meaning less feed is needed to produce the same amount of protein.

2. Versatile Diet:

   - BSF larvae can consume a wide range of organic materials, including food scraps, agricultural residues, and manure. This versatility makes them valuable for waste management and reduces the need for specialized feed.

3. Waste Reduction:

- As larvae consume organic waste, they reduce the volume of waste material by up to 50%. This not only mitigates waste disposal issues but also produces valuable by-products like frass, which can be used as a fertilizer.

4. Low Environmental Impact:

- BSF farming requires significantly less water and land compared to traditional livestock farming. Additionally, the process emits fewer greenhouse gases, contributing to a lower environmental footprint.

Ways the black soldier fly life cycle can be optimized

There are several ways the black soldier fly life cycle can be optimized for large-scale maggot farming operations:

Egg Production Optimization:

Providing female flies with suitable oviposition (egg-laying) sites and substrates can maximize egg production.

Controlling environmental factors like temperature, humidity, and light exposure can enhance egg-laying behavior.

Selecting for fly lines with higher fecundity (egg-laying capacity) through selective breeding.

Larval Growth Acceleration:

Maintaining optimal temperature (25-30°C), moisture, and aeration in the larval rearing substrate.

Providing a nutrient-rich diet, such as a balanced mixture of organic waste materials, to support rapid growth.

Implementing feeding strategies that maximize larval density and minimize competition for resources.

Genetic selection for faster-growing larval lines.

Pupation and Adult Emergence Management:

Designing rearing systems that facilitate the natural migration of mature larvae to pupation sites.

Controlling the environmental cues (e.g., light, humidity) that trigger pupation and adult emergence.

Optimizing the conditions in the pupal stage to ensure high adult fly survival and fertility.

Closed-Loop System Integration:

The integration of the black soldier fly (BSF) life cycle into closed-loop systems represents a transformative approach to sustainable resource management and waste upcycling. By leveraging the remarkable bioconversion capabilities of BSF larvae, these innovative systems can unlock tremendous environmental and economic benefits.

At the core of this closed-loop integration is the ability of BSF larvae to thrive on a diverse array of organic waste streams, including agricultural byproducts, industrial waste, and municipal solid waste. These larvae possess the unique ability to efficiently convert these waste materials into high-value biomass rich in protein, fats, and other essential nutrients.

By incorporating BSF production into a closed-loop system, the waste streams from other agricultural or industrial processes can be seamlessly repurposed as feedstock for the larvae. This integration creates a self-sustaining cycle where the outputs of one system become the inputs for another, minimizing waste and maximizing resource utilization.

For example, in a livestock production facility, the manure and other organic waste can be diverted to BSF larvae cultivation. The larvae, in turn, can be harvested and processed into high-quality animal feed, effectively closing the loop and reducing the reliance on external feed sources. Similarly, in food processing industries, the organic waste generated can be channeled into BSF production, transforming waste into valuable protein and fat-rich feed ingredients.

This closed-loop integration not only optimizes resource efficiency but also delivers significant environmental benefits. By diverting organic waste from landfills or inefficient disposal methods, these systems help mitigate greenhouse gas emissions and reduce the strain on natural resources. Furthermore, the nutrient-dense BSF-based feed can displace the need for land-intensive and water-demanding crop cultivation for animal feed production.

As the global community grapples with the challenges of sustainable food production, waste management, and environmental preservation, the integration of BSF life cycles into closed-loop systems offers a promising and holistic solution. By harnessing the power of nature's bioconversion processes, these innovative systems can contribute to the transition towards a more circular and resilient food and agricultural ecosystem.

Automation and Scaling Technologies:

Developing automated systems for egg collection, larval rearing, harvesting, and post-processing to increase production capacity and efficiency.

Implementing sensor-based monitoring and control systems to optimize environmental conditions and streamline operations.

As the adoption of black soldier fly (BSF) production systems gains momentum, the integration of advanced automation and scaling technologies has become crucial to enhance efficiency, productivity, and cost-effectiveness.

Automated Larval Rearing Systems

One of the key advancements in BSF production is the development of automated larval rearing systems. These systems leverage robotics, sensor technologies, and advanced control algorithms to streamline the feeding, monitoring, and harvesting of BSF larvae. Automated feeder systems can precisely dispense the appropriate amount of organic waste substrates, ensuring optimal nutritional intake and growth conditions for the larvae. Integrated sensor networks can continuously monitor environmental parameters, such as temperature, humidity, and pH, allowing for real-time adjustments to maintain optimal rearing conditions.

Automated Harvesting and Processing

The harvesting and processing of BSF larvae is another area where automation has made significant strides. Robotic systems can be employed to efficiently collect the mature larvae, separating them from the substrate and transporting them for further processing. Automated drying, separation, and packaging systems can handle the post-harvest processing, ensuring consistent product quality and reducing labor-intensive manual tasks.

Modular and Scalable Production Facilities

To accommodate the growing demand for BSF-based products, the development of modular and scalable production facilities has become crucial. These facilities are designed with a modular architecture, allowing for easy expansion and the ability to adapt to changing market conditions or feedstock availability. The use of prefabricated, stackable rearing units enables the rapid deployment of production capacity, reducing the time and cost associated with traditional facility construction.

Data-Driven Optimization and Process Control

Integrating data-driven optimization and process control technologies can further enhance the efficiency and scalability of BSF production systems. Advanced data analytics, machine learning, and predictive modeling can be employed to continuously monitor and optimize key performance indicators, such as growth rates, feed conversion ratios, and product yields. This data-driven approach enables real-time adjustments to the rearing process, ensuring consistent and reliable BSF biomass production.

By embracing these automation and scaling technologies, BSF production systems can achieve increased efficiency, reduced labor costs, and enhanced scalability to meet the growing demand for sustainable protein and nutrient sources. As the industry continues to evolve, the integration of these cutting-edge technologies will be instrumental in driving the widespread adoption and commercialization of BSF-based solutions.

Scaling up rearing facilities and infrastructure to achieve economies of scale in large-scale maggot farming.

By optimizing the various stages of the black soldier fly life cycle, maggot farming operations can increase productivity, consistency, and profitability, making it a more viable and sustainable solution for waste management and feed production.

Practical Applications of Maggot Farming

1. Animal Feed:

   - BSF larvae can be used to produce high-quality feed for livestock, poultry, and aquaculture. Their high protein and fat content make them an ideal ingredient for balanced animal diets.

How to calculate the organic (maggot feed) needed

To calculate the organic matter (maggot feed) needed for Black Soldier Fly (BSF) maggot production, consider the following factors:

1. Maggot production goal: Determine the desired weight or quantity of maggots to be produced.

2. Feed conversion ratio (FCR): BSF maggots can convert 1.5-2.5 times their body weight in feed into biomass. Use an FCR of 2 for calculations.

3. Feed moisture content: BSF maggots thrive on moist feed (50-60% moisture). Adjust the calculation based on the feed's moisture content.

4. Feed density: Estimate the density of the feed material (e.g., 0.5-1.0 g/cm^3).

Calculation steps:

1. Determine the total biomass production goal (maggot weight x FCR).

2. Calculate the required feed weight (biomass production goal / feed conversion ratio).

3. Adjust for feed moisture content (feed weight x moisture content).

4. Calculate the volume of feed needed (feed weight / feed density).

Example:- Maggot production goal: 100 kg

- FCR: 2

- Feed moisture content: 55%

- Feed density: 0.7 g/cm^3

- Biomass production goal: 100 kg x 2 = 200 kg

- Required feed weight: 200 kg / 2 = 100 kg

- Adjusted feed weight (55% moisture): 100 kg x 0.55 = 55 kg

- Volume of feed needed: 55 kg / 0.7 g/cm^3 ≈ 79 cm^3 (or 79 liters)

Please note that these calculations serve as an estimate, and actual feed requirements may vary depending on specific conditions and BSF maggot growth rates.

Management of daily Production

Effective management of daily production in a BSF maggot farm involves:

1. Scheduling: Plan and schedule tasks, such as feeding, monitoring, and harvesting.

2. Feed management: Ensure consistent feed quality and quantity for optimal maggot growth.

3. Environmental control: Monitor and maintain optimal temperature, humidity, and ventilation levels.

4. Maggot monitoring: Regularly check maggot health, growth, and development.

5. Harvesting and processing: Efficiently collect and process maggots for sale or further processing.

6. Cleaning and sanitation: Maintain a clean and sanitary environment to prevent disease and contamination.

7. Record keeping: Accurately track production data, feed consumption, and environmental conditions.

8. Staff management: Train and supervise staff to ensure efficient and effective production practices.

9. Quality control: Implement quality control measures to ensure consistent product quality.

10. Continuous improvement: Regularly evaluate and improve production processes to optimize efficiency and productivity.

Let me break down each aspect of managing daily production in a BSF maggot farm:

1. Feed Management:

    - Prepare feed according to recipes and schedules

    - Monitor feed consumption and adjust as needed

    - Ensure proper feed storage and handling

2. Environmental Control:

    - Monitor and control temperature, humidity, and ventilation

    - Maintain optimal environmental conditions for maggot growth

    - Adjust environmental parameters as needed

3. Maggot Monitoring:

- Regularly check maggot health, growth, and development

- Monitor for signs of disease, stress, or contamination

- Take corrective action if issues arise

4. Harvesting and Processing:

- Schedule and execute harvesting and processing tasks efficiently

- Ensure proper handling and storage of harvested maggots

- Follow proper processing procedures for desired products (e.g., meal, oil, or live maggots)

5. Cleaning and Sanitation:

- Clean and sanitize facilities, equipment, and tools regularly

- Implement biosecurity measures to prevent disease introduction

- Ensure proper waste management and disposal

6. Record Keeping:

- Accurately track production data, feed consumption, and environmental conditions

- Use data to optimize production processes and make informed decisions

- Maintain records for regulatory compliance and quality control

7. Staff Management:

- Train and supervise staff to ensure efficient and effective production practices

- Encourage teamwork and accountability

- Ensure proper staffing levels and scheduling

8. Quality Control:

- Implement quality control measures to ensure consistent

product quality

   - Monitor and adjust processes as needed to meet quality standards

   - Conduct regular quality checks and testing

9. Continuous Improvement:

   - Regularly evaluate and improve production processes to optimize efficiency and productivity

   - Encourage innovation and experimentation

   - Stay up-to-date with industry best practices and new technologies

Focusing on these aspects, you can ensure a well-managed and efficient daily production process in your BSF maggot farm. By implementing these practices, you can ensure a well-managed and efficient daily production process in your BSF maggot farm.

2. Waste Management:

   - By converting organic waste into valuable biomass, maggot farming offers an effective solution for waste reduction. This can be particularly beneficial for municipalities, agricultural operations, and food processing industries.

3. Human Consumption:

   - While less common, BSF larvae have potential as a sustainable protein source for human consumption. They can be processed into protein powders or used in various food products.

Is the BSF maggot Edible for human beings?

Yes, black soldier fly (BSF) maggots are edible and have been used as a food source for humans and animals. Here are some points to consider:

- High in protein: BSF maggots have a high protein content, making them a potential source of nutrition for humans and animals.

- Low carbon footprint: Maggot farming has a lower carbon footprint than traditional livestock farming.

- Food waste reduction: Maggots can consume food waste, reducing the amount of waste sent to landfills.

- Animal feed: Maggots are already used as feed for certain animals, such as reptiles and birds.

- Human consumption: While not widely accepted in Western cultures, BSF maggots have been consumed in some Asian cultures.

- Safety: BSF maggots are considered safe for human consumption, but proper handling and preparation are essential.

- Taste and texture: Described as having a nutty flavor and crunchy texture.

- Commercial production: Some companies already produce BSF maggots for human consumption and animal feed.

4. Industrial Applications:

   - The fats and proteins extracted from BSF larvae can be used in cosmetics, biofuels, and other industrial applications, showcasing their versatility beyond traditional farming.

Conclusion

The science behind maggot farming reveals a highly efficient, sustainable, and versatile practice with immense potential. By understanding the nutritional profile, life cycle, and efficiency mechanisms of BSF larvae, we can better appreciate the benefits and applications of maggot farming. This knowledge sets the foundation for practical implementation, which we will explore in the subsequent chapters. Whether you aim to improve waste management, enhance animal feed production, or venture into innovative industrial uses, maggot farming offers a promising pathway to a sustainable future.

CHAPTER 3: SETTING UP YOUR MAGGOT FARM

Setting up a maggot farm, particularly for Black Soldier Flies (BSF), requires careful planning and attention to detail. This chapter will guide you through the essential steps and considerations involved in establishing your own maggot farming operation. From choosing the right location to designing the infrastructure and selecting suitable substrates, each decision will contribute to the success and efficiency of your farm.

However, if you're interested in starting a BSF larvae production system, here are some general steps for setting up a starter kit:

Research and Planning:

Understand the basics of BSF biology, life cycle, and their role in waste management and protein production. Determine the purpose of your BSF production (e.g., animal feed, bioconversion, composting).

How to Calculate facilities/construction needed

To calculate the facilities/construction needed for BSF maggot production, consider the following factors:

1. _Production capacity_: Desired maggot production rate (e.g., kg/day or ton/month).

2. _Maggot growth stages_: Separate areas for egg laying, larval growth, and pupation.

3. _Space requirements_:

   - Egg laying: 1-2 square meters (10-20 sq ft) per ton of maggots produced.

   - Larval growth: 5-10 square meters (50-100 sq ft) per ton of maggots produced.

   - Pupation: 2-5 square meters (20-50 sq ft) per ton of maggots produced.

4. _Building design_:

   - Climate control: temperature, humidity, and ventilation.

   - Lighting: appropriate intensity and spectrum for BSF development.

   - Drainage and sanitation: easy cleaning and waste management.

5. _Equipment and infrastructure_:

   - Breeding tanks or trays.

   - Larval rearing trays or containers.

   - Pupation boxes or containers.

   - Aeration and ventilation systems.

   - Drainage and water management systems.

6. _Utilities and services_:

   - Water supply.

   - Electricity.

   - Waste management.

Calculation steps:

1. Determine the total production capacity (kg/day or ton/month).

2. Calculate the space requirements for each growth stage (egg laying, larval growth, and pupation).

3. Add the space requirements for each stage to determine the total facility size.

4. Consider the building design and equipment needs based on the calculated space requirements.

5. Estimate the utilities and services needed to support the facility.

Example:

- Production capacity: 1 ton of maggots per month.

- Egg laying: 2 square meters (20 sq ft) per ton of maggots produced.

- Larval growth: 5 square meters (50 sq ft) per ton of maggots produced.

- Pupation: 2 square meters (20 sq ft) per ton of maggots produced.

- Total facility size: 2 + 5 + 2 = 9 square meters (90 sq ft).

Please note that these calculations serve as an estimate, and actual facility requirements may vary depending on specific conditions and BSF maggot growth rates. It's recommended to consult with experts in BSF maggot production and facility design to get more accurate calculations and guidance.

Select a Suitable Location:

Choose an area with proper ventilation, temperature control, and protection from pests.

Consider indoor or outdoor setups based on your resources and climate.

Build or Acquire Containers:

You'll need containers for larvae breeding and substrate.

Start with smaller containers for your starter kit.

Prepare Substrate:

BSF larvae thrive on organic waste (e.g., kitchen scraps, manure, food waste).

Mix the substrate with bedding material (e.g., wood shavings, coconut coir).

Introduce BSF Eggs or Larvae:

Obtain BSF eggs or larvae from a reliable source.

Place them in the substrate.

Monitor and Maintain:

Regularly check temperature, humidity, and substrate moisture.

Provide adequate food (organic waste) for the larvae.

Harvest and Expand:

Once larvae mature, harvest them for your intended purpose.

Consider scaling up by transferring larvae to larger containers.

Remember that BSF larvae production requires attention to hygiene, proper feeding, and environmental conditions.

Choosing the Right Location

The location of your maggot farm plays a crucial role in its productivity and sustainability. Consider the following factors when selecting a site:

1. Climate and Temperature:

   - BSF larvae thrive in warm temperatures, ideally between 25°C to 30°C (77°F to 86°F). Choose a location where temperatures remain within this range for most of the year.

   - Extreme temperatures, both hot and cold, can impact larval development and overall farm productivity.

2. Sunlight Exposure:

- Adequate sunlight is beneficial for BSF larvae and their development. Choose a location that receives sufficient natural light or plan for artificial lighting if needed, especially in colder climates or during winter months.

3. Accessibility and Security:

- Ensure the farm site is easily accessible for deliveries and waste collection. It should also be secure to prevent unauthorized access and protect against potential pests or predators.

4. Proximity to Organic Waste Sources:

- To minimize transportation costs and logistical challenges, consider locating your farm near sources of organic waste, such as food processing facilities, markets, or agricultural operations.

Designing the Infrastructure

The infrastructure of your maggot farm should be designed to optimize space, facilitate waste management, and provide a conducive environment for BSF larvae:

Building orientation plays a significant role in the cultivation of Black Soldier Fly (BSF), particularly when considering optimal conditions for larvae growth and overall farm efficiency. Here are key considerations and discussions on building orientation for BSF cultivation:

Importance of Building Orientation

1. Sunlight Exposure:

- South-Facing Orientation: In many regions, a south-facing orientation is preferred for BSF farms. This orientation maximizes exposure to sunlight throughout the day, which is beneficial for maintaining optimal temperatures and promoting larval activity.

- Natural Light: Adequate natural light is crucial for BSF

larvae development. South-facing buildings receive more sunlight during the day, especially in northern hemispheres, which helps regulate temperature and supports photosynthesis in any plants that might be used for larval substrate.

2. Temperature Regulation:

- Temperature Control: BSF larvae thrive in warm temperatures, typically between 25°C to 30°C (77°F to 86°F). Proper building orientation can help regulate internal temperatures, reducing the need for artificial heating or cooling systems.

- Avoiding Overheating: While sunlight is beneficial, excessive heat buildup inside the building can be detrimental. Proper ventilation and shading may be necessary to prevent overheating during hot weather.

3. Ventilation and Airflow:

- Natural Ventilation: Orienting the building to take advantage of prevailing winds can enhance natural ventilation within the farming area. Good airflow helps maintain optimal humidity levels and prevents stagnant air, which can lead to mold or bacterial growth.

- Cross-Ventilation: Designing the building to allow for cross-ventilation can further improve airflow, ensuring a healthy environment for BSF larvae.

4. Protection from Elements:

- Wind and Rain: Consider how prevailing winds and rain patterns might affect the building and larval habitats. Proper orientation can provide natural protection against harsh weather conditions, reducing potential damage or disruption to farm operations.

- Sheltered Areas: Buildings can be oriented to create sheltered areas that protect larvae and farming equipment from direct exposure to wind or heavy rainfall.

5. Accessibility and Efficiency:

- Ease of Access: Orienting the building for convenient access to waste collection points, storage areas, and operational facilities improves overall farm efficiency.

- Workflow Optimization: Thoughtful orientation can streamline workflow processes, from waste handling to larval harvesting, contributing to operational efficiency and reducing labor costs.

Practical Considerations

- Local Climate Conditions: Adapt building orientation based on local climate patterns, such as seasonal temperature variations and prevailing wind directions.

- Building Design: Integrate passive design strategies, such as thermal mass, insulation, and shading devices, to optimize indoor conditions and reduce energy consumption.

- Site-Specific Factors: Assess site-specific factors, including topography, neighboring structures, and land use regulations, that may influence building orientation decisions.

Let me provide more details on the key aspects of building and operating a successful maggotery for black soldier fly (BSF) larvae rearing:

Site Selection:

Choosing the right location is crucial, as it can impact the overall success and sustainability of the operation.

In addition to being away from human residences, the site should have access to reliable sources of organic waste feedstock, which is the primary food source for the BSF larvae.

Proximity to water sources, such as a well or a river, can be beneficial for managing moisture levels and facilitating cleaning.

The site should have good drainage to prevent waterlogging and potential breeding grounds for mosquitoes or other pests.

Considering the prevailing wind direction can help mitigate the spread of odors to nearby areas.

Housing Design:

The building should be designed to optimize airflow and temperature regulation, as BSF larvae thrive in specific environmental conditions.

Proper ventilation, either through natural airflow or mechanical systems, is essential to prevent the buildup of moisture, heat, and odors.

The use of materials that are easy to clean and disinfect, such as smooth surfaces and non-porous materials, can help maintain hygiene and prevent the spread of diseases.

Incorporating features like shading, insulation, or climate control systems can help regulate the temperature and humidity within the maggotery, ensuring optimal conditions for the larvae.

Housing:

The building should be oriented in an east-west direction to minimize the effects of direct sunlight on the substrates.

The building should have openings at the top and sides to ensure proper ventilation.

The roof can be made from corrugated iron sheets or thatched material.

The dimensions of the building should be:

3 meters from the top to the feet of the building

3.2 meters from the floor to the ridge of the building

Short walls (maximum of 3 block courses) and open-sided to enable aeration.

These considerations are crucial for designing an effective and efficient maggotery facility. The site selection and housing specifications aim to mitigate the potential odor issues, optimize the environmental conditions, and promote proper airflow and ventilation for the rearing of black soldier fly larvae.

Accessibility for maintenance, harvesting, and waste management should also be considered in the layout and design of the facility.

Substrate and Feeding Management:

The type and composition of the organic waste substrate used to feed the BSF larvae can have a significant impact on their growth and development.

Diversifying the feedstock, such as using a mixture of food waste, agricultural residues, and animal manure, can provide a balanced nutritional profile for the larvae.

Proper moisture management, aeration, and regular substrate replenishment are essential to maintain optimal conditions for the larvae.

Monitoring and adjusting the feeding rates and substrate composition based on the larvae's growth and development can help optimize the rearing process.

Biosecurity and Disease Prevention:

Implementing strict biosecurity measures, such as sanitization protocols, quarantine procedures, and pest management strategies, can help prevent the introduction and spread of diseases within the maggotery.

Regular monitoring and early detection of potential disease or pest outbreaks are crucial for timely intervention and mitigation.

Consulting with experts, such as entomologists or veterinarians, can provide guidance on effective disease management and control strategies.

By considering these aspects in the design, construction, and operation of a maggotery, BSF producers can establish a robust and sustainable rearing system that maximizes the health, growth, and productivity of the black soldier fly larvae.

Bramell's five freedoms

The Five Freedoms Approach for Animal Welfare in Black Soldier Fly Production. The Five Freedoms, as proposed by the British Farm Animal Welfare Council (FAWC) and championed by Dr. Brambell, provide a comprehensive framework for ensuring the welfare of animals in various production systems, including the cultivation of black soldier fly (BSF) larvae.

1. Freedom from Hunger and Thirst

Ensuring that BSF larvae have access to a sufficient and appropriate supply of nutritious feed is crucial. This includes providing a diverse range of organic waste substrates that meet the larvae's nutritional requirements for growth and development.

2. Freedom from Discomfort

Maintaining optimal environmental conditions, such as temperature, humidity, and substrate depth, is essential to prevent discomfort and ensure the well-being of the BSF larvae. Appropriate housing and rearing systems should be designed to minimize stress and allow the larvae to exhibit their natural behaviors.

3. Freedom from Pain, Injury, or Disease

Implementing robust biosecurity measures, disease prevention protocols, and prompt veterinary care can help safeguard the BSF larvae from any potential sources of pain, injury, or disease. Regular monitoring and proactive interventions can minimize the risk of health-related issues.

4. Freedom to Express Normal Behavior

Providing the BSF larvae with a suitable substrate and rearing environment that allows them to exhibit their natural behaviors, such as burrowing, feeding, and pupation, is crucial for their overall well-being. This can include maintaining appropriate substrate depth, texture, and organic matter composition.

5. Freedom from Fear and Distress

Minimizing any sources of fear, anxiety, or distress for the BSF larvae is essential. This may involve preventing sudden changes in environmental conditions, avoiding excessive handling or disruptions, and ensuring a calm and stress-free production environment.

By adhering to the principles of the Five Freedoms, BSF production systems can ensure the welfare of the larvae, promoting their health, growth, and natural behaviors. This not only upholds ethical considerations but can also contribute to the overall productivity and sustainability of the BSF production process.

Incorporating the Five Freedoms approach into the design, management, and monitoring of BSF production systems can serve as a guiding framework to prioritize animal welfare and align with best practices in sustainable and responsible insect farming.

The Five Freedoms framework provides a robust foundation for optimizing the production of black soldier fly (BSF) larvae, ensuring their well-being while maximizing the efficiency and sustainability of the system. Here's how each of the five freedoms can be applied to BSF production:

Freedom from Hunger and Thirst:

Provide a diverse range of organic waste substrates that meet the nutritional requirements of the BSF larvae, including appropriate protein, carbohydrate, and fat content.

Implement a feeding regime that ensures the larvae have access to a sufficient and consistent supply of feed.

Monitor and maintain the moisture content of the substrate to prevent dehydration.

Freedom from Discomfort:

Maintain optimal temperature and humidity levels within the rearing environment to provide a comfortable and suitable habitat for the larvae.

Ensure appropriate substrate depth and texture to allow the larvae to burrow and move freely.

Minimize disruptions and disturbances that could cause stress or discomfort to the larvae.

Freedom from Pain, Injury, or Disease:

Develop and implement comprehensive biosecurity protocols to prevent the introduction and spread of pathogens.

Closely monitor the larvae for signs of illness or injury and provide prompt veterinary care if necessary.

Optimize the rearing conditions to minimize the risk of disease outbreaks or physical trauma.

Freedom to Express Normal Behavior:

Design the rearing systems to mimic the natural habitat and life cycle of the BSF, allowing the larvae to exhibit their innate behaviors, such as burrowing, feeding, and pupation.

Provide a suitable substrate composition and depth to facilitate the larvae's natural behaviors.

Minimize any disruptions or interventions that could hinder the expression of the larvae's natural behaviors.

Freedom from Fear and Distress:

Maintain a calm and low-stress production environment, minimizing sudden changes, loud noises, and excessive handling.

Implement gentle handling techniques and avoid any practices that may induce fear or anxiety in the larvae.

Monitor the larvae for signs of stress and make necessary adjustments to the rearing conditions.

By applying the Five Freedoms approach, BSF production systems can optimize the well-being of the larvae, leading to improved growth rates, feed conversion efficiency, and overall productivity. This holistic approach not only benefits the larvae but also contributes to the long-term sustainability and viability of the BSF production process.

Integrating the Five Freedoms into the design, management, and continuous improvement of BSF production systems can help ensure ethical and responsible insect farming practices, aligning with the growing societal demand for sustainable and animal-friendly food and agriculture systems. Building orientation is a critical consideration in BSF cultivation, influencing sunlight exposure, temperature regulation, ventilation, and overall operational efficiency. By strategically orienting the building to align with local climate conditions and operational needs, BSF farmers can create an optimal environment for larval growth and maximize the sustainability and productivity of their farming operations.

1. Farm Layout:

   - Plan the layout of your farm based on the scale of operation and available space. Vertical farming systems or racks can maximize space utilization, allowing for efficient larval growth and harvest.

2. Containers and Enclosures:

   - Use containers or enclosures that are durable, easy to clean, and resistant to weather conditions. Plastic bins or trays with adequate drainage are commonly used for rearing BSF larvae.

3. Ventilation and Airflow:

   - Ensure proper ventilation and airflow within the farming area to maintain optimal conditions for larvae. Good airflow helps regulate temperature and humidity, preventing moisture buildup and potential issues like mold.

4. Lighting Options:

- Natural sunlight is ideal, but supplementary lighting may be necessary, especially in indoor or shaded environments. Consider LED grow lights that emit wavelengths beneficial for larval growth and development.

Selecting Suitable Substrates

Choosing the right substrate or organic waste material is essential for feeding BSF larvae and promoting their growth:

1. Organic Waste Sources:

   - BSF larvae can consume a wide range of organic materials, including fruit and vegetable scraps, brewery waste, coffee grounds, and manure.

   - Ensure the substrates are free from contaminants or toxins that could harm larvae or affect the quality of the resulting biomass.

2. Preparing Substrates:

   - Depending on the type of waste, substrates may need to be processed or shredded to facilitate larval feeding and digestion. Uniform particle size and proper moisture content are critical for optimal consumption.

3. Nutrient Balance:

   - Provide a balanced diet for BSF larvae by mixing different types of organic waste. This helps ensure larvae receive essential nutrients, including proteins, fats, carbohydrates, vitamins, and minerals necessary for growth.

Implementing Biosecurity Measures

Implementing biosecurity measures in your maggot farm is crucial to protect the health and productivity of your insects. Some key steps to implementing biosecurity measures include:

- Controlling access to your farm to minimize the risk of introducing pests or diseases.
- Quarantining new insects or feed sources before

introducing them to your farm.
- Regularly cleaning and disinfecting equipment and facilities to prevent the spread of contaminants.
- Monitoring and maintaining the health and hygiene of your insects.
- Implementing strict hygiene practices for yourself and farm workers to prevent the spread of diseases.

In 2010, the World Health Organization (WHO) issued an informational note defining biosecurity as a comprehensive approach to assessing and mitigating risks to human, animal, and plant life and health, including risks to the environment. In another document, WHO articulated that the primary objective of biosecurity is to strengthen the capacity to safeguard human health, agricultural systems, and the associated communities and industries. The overarching aim is to prevent, control, and manage risks to life and health within specific biosecurity domains as deemed appropriate. Biosecurity is essential to prevent the introduction and spread of pests, diseases, or contaminants within your maggot farm:

1. Quarantine and Monitoring:

   - Implement quarantine measures for incoming organic waste to prevent potential contaminants or pests from entering your farm.

   - Regularly monitor larval health and farm conditions to detect any signs of disease or stress early on.

2. Waste Management Protocols:

   - Establish proper waste management protocols to handle and dispose of spent substrates or larvae that do not meet quality standards.

   - Composting or using frass (larval excrement) as fertilizer can be part of a sustainable waste management strategy.

Conclusion

Setting up a maggot farm requires careful planning and attention to detail across various aspects, from choosing the right location and designing infrastructure to selecting suitable substrates and implementing biosecurity measures. By following these guidelines and adapting them to your specific circumstances, you can create a productive and sustainable environment for rearing Black Soldier Fly larvae. In the next chapter, we will explore the practical aspects of breeding and managing BSF larvae, essential for optimizing productivity and achieving successful harvests in your maggot farming venture.

CHAPTER 4: BREEDING AND MANAGING BLACK SOLDIER FLY LARVAE

Breeding and managing Black Soldier Fly (BSF) larvae effectively is essential for maximizing biomass production and maintaining a healthy farming operation.

This chapter explores the lifecycle of BSF larvae, optimal breeding practices, feeding protocols, and management techniques to ensure sustainable and productive outcomes.

Understanding the Lifecycle of BSF

Efficient breeding and management of black soldier fly (BSF) larvae are crucial for the successful and sustainable production of BSF-based products. Here's an overview of the key aspects involved in breeding and managing BSF larvae:

Breeding and Reproduction:

Establish a breeding colony of adult BSF with a balanced sex ratio (around 1:1) to ensure successful mating and egg production. The sex ratio of male to female Black Soldier Flies (BSF) among eggs laid is typically around 1:1, meaning that approximately half of the eggs will hatch into males and half will hatch into females. However, some studies have reported a slightly biased sex ratio, with a range of:

- 1:1.2 to 1:1.5 (males:females)
- 48% males and 52% females
- 45% males and 55% females

This slight bias towards females may be due to various factors, such as:

- Genetic influences
- Environmental conditions
- Nutrition and diet

Identifying male and female Black Soldier Flies (BSF)

Identifying male and female Black Soldier Flies (BSF) requires a close examination of their physical characteristics. Here's a step-by-step guide:

Male BSF (Hermetia illucens):

1. Smaller size: Males are slightly smaller than females.

2. Brighter color: Males have a more vibrant, metallic blue-black color.

3. Longer antennae: Males have longer antennae with a more pronounced club shape at the tip.

4. Narrower abdomen: The abdomen is slender and tapering.

5. Modified wings: Males have slightly modified wings with a more rounded tip.

Female BSF (Hermetia illucens):

1. Larger size: Females are slightly larger than males.

2. Duller color: Females have a duller, more matte black color.

3. Shorter antennae: Females have shorter antennae with a less

pronounced club shape.

4. Wider abdomen: The abdomen is broader and more rounded.

5. Unmodified wings: Females have unmodified wings with a more pointed tip.

Additional tips:

- Males tend to have a more rounded thorax (middle segment) than females.

- Females often have a visible ovipositor (egg-laying structure) at the end of their abdomen.

- It's important to note that sexing BSF can be challenging, especially for inexperienced individuals. It's recommended to consult with an expert or use specialized resources for accurate identification.

Remember, accurate sexing is crucial for breeding and reproduction purposes in BSF farming.It's important to note that the sex ratio can vary depending on the specific breeding stock, rearing conditions, and management practices. BSF farmers and breeders often aim to maintain a balanced sex ratio to ensure optimal reproduction and egg production.

Mating and oviposition substrate

Provide the adult flies with a suitable mating and oviposition substrate, such as a shallow container with a rough surface and organic matter. Here are the key characteristics of a suitable mating and oviposition substrate for BSF:

Substrate Material:

A combination of organic materials, such as decomposing plant matter, dried leaves, straw, or aged animal bedding, can provide an ideal substrate.

The organic matter not only serves as a nutritional source for the adult flies but also emits the chemical cues that stimulate mating and egg-laying behaviors.

Surface Texture:

The substrate should have a rough or uneven surface to provide traction and stability for the adult flies during mating.

Materials like burlap, corrugated cardboard, or coarse mesh can create the desired textured surface.

Moisture Content:

The moisture content of the substrate should be maintained between 60-80% to provide a suitable microclimate for the adult flies and ensure the eggs remain hydrated during incubation.

Periodic misting or adding a small amount of water may be necessary to maintain the optimal moisture level.

Depth and Accessibility:

The substrate should be placed in a shallow container or on a sloped surface to allow easy access for the adult flies.

A depth of 5-10 cm is generally suitable, as it provides enough space for the flies to move and deposit their eggs without being too deep.

Enclosure and Ventilation:

The substrate should be contained within a dedicated enclosure or container, such as a mesh-covered tray or a plastic bin, to prevent the escape of adult flies and facilitate egg collection.

Adequate ventilation should be provided to ensure proper air circulation and maintain the desired environmental conditions.

Lighting and Temperature:

Appropriate lighting conditions, such as a combination of natural and artificial light, can help stimulate the flies' mating and oviposition behaviors.

The temperature in the breeding area should be maintained between 25-30°C, as this range is optimal for the adult flies' activities.

By providing a suitable mating and oviposition substrate that meets the above criteria, BSF producers can create an environment that encourages successful mating, egg-laying, and the subsequent development of a healthy larval population. Regular monitoring and adjustments to the substrate's characteristics can further optimize the breeding process and ensure a consistent supply of BSF larvae for various applications.

Maintain optimal environmental conditions, including temperature (25-30°C), humidity (60-80%), and lighting regime, to stimulate mating and egg-laying behavior.

Collect the egg clusters laid by the female flies and transfer them to a dedicated hatching environment.

Monitor the egg development and ensure the newly hatched larvae have access to a suitable rearing substrate.

Larval Rearing:

Prepare the rearing substrate by selecting appropriate organic waste materials, such as food waste, agricultural byproducts, or animal manure, ensuring a balanced nutritional profile.

Maintain the substrate's moisture content and pH within the optimal range for BSF larval growth and development.

Distribute the newly hatched larvae onto the rearing substrate, ensuring an appropriate stocking density to prevent overcrowding.

Monitor the larval growth and development, adjusting the feeding regime, environmental conditions, and substrate management as needed.

Implement effective harvesting methods, such as using vibrations or light to encourage the mature larvae to migrate from the substrate for collection.

Larval Management:

Establish a comprehensive biosecurity protocol to prevent the introduction and spread of pathogens, including regular cleaning and disinfection of the rearing environment.

Closely monitor the larvae for any signs of disease or stress and take prompt corrective actions, such as adjusting environmental parameters or providing appropriate medical interventions. Diseases and Pathogens Affecting Black Soldier Fly (BSF) Larvae

Diseases and pathogens that can affect BSF larvae:

Black soldier fly (BSF) larvae are generally considered a robust and resilient species, but like any living organism, they can be susceptible to various diseases and pathogens. Understanding the common diseases and implementing effective preventive measures are crucial for maintaining the health and productivity of a BSF rearing operation.

Bacterial Infections:

Bacterial pathogens, such as Bacillus cereus and Serratia marcescens, can cause septicemia (blood infections) and digestive system issues in BSF larvae.

Symptoms of bacterial infections in BSF larvae may include:

a. Reduced feeding and lethargy:

Affected larvae may exhibit a noticeable decrease in feeding activity and appear sluggish or lethargic.

b. Discoloration and abnormal appearance:

The larvae may display discoloration, such as darkening or reddish hues, on their body segments.

In advanced cases, the larvae may develop a shriveled or deformed appearance.

c. Liquefaction and septicemia:

Severe bacterial infections can lead to the liquefaction of the larval body, resulting in a watery, decomposed appearance.

Septicemia, or blood infections, may cause the larvae to have a reddish or purplish coloration.

d. Altered behavior:

Infected larvae may exhibit abnormal behaviors, such as moving away from the feeding substrate or attempting to escape the rearing environment.

Fungal Infections:

Fungal diseases, particularly those caused by Beauveria bassiana and Aspergillus spp., can lead to mycosis (fungal infections) in BSF larvae.

Affected larvae may exhibit sluggishness, discoloration, and the presence of visible fungal growth on the body.

Viral Infections:

Viruses, such as the black soldier fly iridescent virus (BSFIV), can cause viral diseases in BSF larvae.

Symptoms may include reduced growth, developmental abnormalities, and increased mortality.

Parasitic Infestations:

Parasites, like nematodes or mites, can infect BSF larvae and compromise their health and performance.

Parasitic infestations may result in stunted growth, reduced feeding, and increased susceptibility to secondary infections.

To prevent and manage disease outbreaks in a BSF rearing operation, it is essential to implement the following strategies:

Visible fungal growth:

The presence of white, cotton-like or powdery fungal growth on the surface of the larvae is a clear indication of a fungal infection.

b. Discoloration and stunted growth:

Affected larvae may appear discolored, often with a greenish, brownish, or blackish hue.

Fungal infections can also lead to stunted growth and developmental abnormalities in the larvae.

c. Reduced feeding and activity:

Infected larvae may exhibit a decreased appetite and reduced feeding activity, as well as lethargy and sluggishness.

d. Premature mortality:

In severe cases, fungal infections can lead to increased mortality rates among the BSF larvae.

It's important to note that the specific symptoms may vary depending on the type of bacterial or fungal pathogen involved, the severity of the infection, and the overall health and environmental conditions of the BSF rearing system.

Early detection and prompt intervention are crucial to mitigate the spread of these diseases and maintain the overall health and productivity of the BSF larvae. Regular monitoring, proper biosecurity measures, and consultation with experts can help BSF producers effectively manage and prevent these disease-related challenges.

Maintain Strict Biosecurity:

Regularly clean and disinfect the rearing environment, equipment, and tools used in the BSF production process.

Implement strict protocols for the introduction of new BSF stocks or organic substrates to prevent the introduction of pathogens.

Optimize Environmental Conditions:

Maintain appropriate temperature, humidity, and ventilation levels to create an environment that is less favorable for the proliferation of pathogens.

Ensure the organic substrate used for rearing larvae is of high quality and free from contaminants.

Monitor and Early Detection:

Regularly inspect the BSF larvae for any signs of disease, such as behavioral changes, discoloration, or abnormal growth.

Implement rapid diagnostic techniques, such as microscopic examination or molecular testing, to identify the causative agents.

Implement Appropriate Interventions:

If a disease outbreak is detected, isolate the affected larvae and consult with veterinary or entomological experts to determine the appropriate course of action.

Explore the use of targeted antimicrobial treatments, biological control agents, or other interventions, as recommended by experts, to mitigate the disease and prevent its spread.

By implementing a comprehensive disease management strategy, BSF producers can maintain the health and productivity of their BSF rearing operations, ensuring the consistent supply of high-quality BSF larvae for various applications.

Implement measures to minimize the risk of predation or competition from other organisms that may threaten the BSF larvae.

Maintain detailed records of the breeding, rearing, and harvesting processes to facilitate continuous improvement and optimization.

By focusing on efficient breeding, rearing, and management practices, BSF production systems can achieve optimal larval growth, high survival rates, and consistent biomass yields. This, in turn, contributes to the overall sustainability, scalability, and profitability of the BSF production process, enabling the industry to meet the growing demand for innovative and eco-friendly protein sources.

The lifecycle of BSF consists of distinct stages, each requiring specific conditions and management practices:

1. Egg Stage:

   - Egg Deposition: Female BSF deposit eggs near suitable organic waste materials. Eggs are typically laid in clusters and hatch within a few days under optimal conditions.

   - Optimal Conditions: Maintain temperatures around 25°C to 30°C (77°F to 86°F) and provide adequate humidity for optimal egg development.

2. Larval Stage:

   - Feeding and Growth: BSF larvae emerge from eggs and begin feeding voraciously on organic waste. They require high-protein substrates for rapid growth and development.

   - Feeding Schedule: Provide fresh organic waste regularly to sustain larval growth. Adjust feeding rates based on larval density and substrate availability.

3. Pupal Stage:

   - Metamorphosis: Fully grown larvae enter the pupal stage, where they undergo metamorphosis into adult flies.

   - Environmental Conditions: Pupae require a dry, secure environment with adequate airflow to complete development. Monitor humidity levels to prevent mold and ensure pupal health.

4. Adult Stage:

   - Reproduction: Adult BSF emerge from pupae and focus on mating and egg laying. They do not feed during this stage and have a short lifespan.

   - Egg Laying Strategy: Females lay eggs in suitable locations near organic waste sources to ensure larvae have immediate access to food upon hatching.

Breeding Practices

Effective breeding practices are crucial for maintaining healthy BSF populations and optimizing larval production:

1. Stock Management:

   - Healthy Stock: Start with healthy BSF cultures from reputable sources to establish a robust breeding colony.

   - Genetic Diversity: Maintain genetic diversity within the colony to enhance adaptability and resilience to environmental changes. Maintaining genetic diversity within a maggot farming colony is crucial to enhance the adaptability and resilience of the population over time. Here are some key strategies to achieve this:

Start with a Diverse Founder Population:

- Begin your maggot farming operation with a large and genetically diverse starter colony.
- Obtain breeding stock from multiple sources or locations to introduce greater genetic variability.
- Avoid using a small number of individuals as the founding population, as this can lead to inbreeding and loss of diversity.

Implement Rotational Breeding:

- Establish a system of rotational breeding where you regularly introduce new individuals into the breeding pool.
- Avoid breeding closely related individuals, as this can lead to inbreeding depression and reduced fitness.
- Introduce new genetic material from external sources, such as wild-caught specimens or other farming operations, on a regular basis.

Manage Population Size:

- Maintain a sufficiently large population size to allow for genetic recombination and minimize the effects of genetic drift.
- Avoid overcrowding, as this can increase competition

and lead to a higher rate of genetic bottlenecks.
- Use appropriate stocking densities and provide ample resources to support a thriving, diverse colony.

Selective Breeding:

- Selectively breed individuals with desirable traits, such as faster growth, higher feed conversion efficiency, or increased tolerance to environmental stressors.
- However, be cautious not to over-select for specific traits, as this can reduce the overall genetic diversity of the population.
- Maintain a balance between selective breeding and preserving genetic diversity.

Cryopreservation:

Cryopreservation is the process of preserving biological materials, such as cells, tissues, or organisms, by cooling them to very low temperatures, typically below -130°C (-202°F), and storing them in a state of suspended animation. This technique is used to maintain the viability and genetic integrity of the preserved samples for extended periods of time.

- Consider cryopreserving a portion of the colony as a backup to maintain genetic diversity.
- Periodically sample and cryogenically store genetic material (e.g., eggs, pupae) to create a genetic reserve.

This can help mitigate the risk of accidental loss of genetic diversity due to unexpected events or population crashes.

Documentation and Monitoring:

- Maintain detailed records of the breeding history, lineages, and genetic characteristics of your maggot colony.
- Regularly monitor the genetic diversity of the population using appropriate genetic analysis techniques, such as DNA sequencing or molecular

markers.

Use this information to guide your breeding and management decisions to maintain a healthy, diverse colony.

By implementing these strategies, you can help ensure that your maggot farming operation maintains a genetically diverse and adaptable population, better equipped to withstand environmental changes and challenges over time.

2. Monitoring and Maintenance:

   - Regular Monitoring: Observe larval behavior, growth rates, and pupation patterns regularly to detect any signs of stress or disease.

   - Hygiene Practices: Maintain clean rearing containers and equipment to prevent contamination and ensure larval health.

3. Environmental Control:

   - Temperature and Humidity: Monitor and control environmental conditions to optimize larval development and pupation.

   - Lighting: Provide adequate natural or artificial lighting to support larval activity and ensure proper circadian rhythms.

Feeding Protocols

Proper feeding protocols are essential for ensuring BSF larvae receive adequate nutrition and promoting optimal growth:

1. Substrate Selection:

   - High-Protein Sources: Use diverse organic waste materials rich in proteins, fats, and essential nutrients. Examples include fruit and vegetable scraps, brewery waste, and agricultural residues.

   - Preparation: Process substrates as needed (e.g., chopping, blending) to facilitate larval feeding and digestion.

2. Feeding Frequency:

- Regular Feeding: Provide fresh organic waste regularly to maintain larval activity and prevent starvation.

- Adjustment: Adjust feeding rates based on larval density, substrate availability, and seasonal variations in waste production.

Management Techniques

Effective management techniques contribute to the overall success and sustainability of BSF larval farming:

1. Waste Management:

- Efficient Utilization: BSF larvae efficiently convert organic waste into biomass. Implement effective waste management strategies to maximize resource utilization and minimize environmental impact.

- Composting: Use larval frass (excrement) as a nutrient-rich compost for soil improvement or organic farming practices.

2. Harvesting and Processing:

- Harvest Timing: Time larval harvesting based on developmental stage and desired biomass quality. Harvest larvae before they enter the pupal stage to maximize biomass yield.

- Processing: Dry harvested larvae to produce BSF meal or process them into other value-added products for animal feed or industrial applications.

Conclusion

Breeding and managing Black Soldier Fly larvae requires a holistic approach encompassing lifecycle understanding, breeding practices, feeding protocols, and effective management techniques. By implementing these strategies, BSF farmers can optimize larval production, maintain farm sustainability, and harness the full potential of BSF larvae as a valuable resource in agriculture, waste management, and sustainable protein production. In the next chapter, we will delve into Harvesting ,Processing and future trends in BSF farming, exploring new opportunities and advancements in this rapidly evolving field.

CHAPTER 5: HARVESTING AND PROCESSING

In this chapter, we delve into the essential processes of harvesting BSF maggots, drying and processing them into meal, and efficient storage and packaging techniques to maintain quality and maximize usability.

Harvesting Techniques

1. Timing and Efficiency:

   - Determine the optimal harvesting time based on larval development stage and biomass accumulation.

   - Employ methods such as sieving, mechanical separation, or manual collection to gather mature larvae efficiently.

2. Separation and Cleaning:

   - Separate larvae from residual substrate and waste materials using screens or sieves.

   - Rinse larvae gently to remove any remaining substrate or debris, ensuring cleanliness and product quality.

Drying and Processing Maggots into Meal

1. Drying Methods:

   - Choose suitable drying methods based on scale and resources, including solar drying, oven drying, or freeze-drying.

- Control temperature and airflow during drying to preserve nutritional content and minimize moisture content.

2. Grinding and Milling:

- Grind dried larvae into a fine meal using mills or grinders to achieve consistent particle size.

- Consider additional processing steps, such as sieving, to refine meal texture and remove any remaining particles.

Storage and Packaging

1. Storage Conditions:

- Store BSF larvae meal in airtight containers or bags to prevent exposure to moisture and contaminants.

- Maintain cool, dry storage conditions to extend shelf life and preserve nutritional integrity.

2. Packaging Considerations:

- Use food-grade packaging materials that are durable and protect against light and oxygen exposure.

- Label packages with essential information, including product name, batch number, nutritional content, and storage instructions.

Innovative Applications and Future Trends in BSF Farming

Black Soldier Fly (BSF) farming continues to evolve with advancements in technology, research, and innovative applications. This chapter explores emerging trends, potential applications beyond traditional uses, and the future trajectory of BSF farming in sustainable agriculture and environmental stewardship.

Emerging Applications of BSF Products

BSF larvae and their by-products offer diverse applications beyond traditional animal feed production:

1. Human Consumption:

- Edible Insects: BSF larvae are gaining traction as a nutritious and sustainable protein source for human consumption. Processed into protein powders or incorporated into food products, BSF larvae provide a viable alternative to traditional protein sources.

2. Biodegradation and Waste Management:

- Organic Waste Conversion: BSF larvae excel at converting organic waste into biomass. Expanding BSF farming for waste management purposes can reduce landfill volumes and mitigate environmental pollution.

- Biodegradable Plastics: Research explores BSF larvae's potential to biodegrade organic plastics, contributing to sustainable waste management solutions.

3. Animal Feed and Aquaculture:

- High-Quality Protein: BSF larvae are rich in protein, essential amino acids, and fats, making them an excellent ingredient in animal feed formulations.

- Aquaculture Feeds: BSF larvae-based feeds enhance growth rates and nutritional profiles in aquaculture species, reducing reliance on conventional feed sources like fishmeal.

Technological Advancements in BSF Farming

Innovative technologies are enhancing efficiency and scalability in BSF farming:

1. Automated Farming Systems:

- Precision Farming: Automated systems monitor environmental conditions, feed delivery, and larval health, optimizing production efficiency and reducing labor requirements.

- IoT Integration: Internet of Things (IoT) technologies enable real-time data monitoring and remote management of BSF farms, enhancing operational control and decision-making.

2. Genetic Improvement:

- Selective Breeding: Research focuses on selective breeding programs to enhance BSF traits such as growth rate, nutritional content, and resistance to stressors.

- Genetic Engineering: Genetic manipulation aims to enhance specific traits in BSF larvae, potentially increasing their suitability for diverse applications.

Environmental and Economic Benefits

BSF farming offers significant environmental and economic advantages:

1. Resource Efficiency:

- Water and Land Use: BSF farming requires minimal water and land compared to traditional livestock farming, reducing environmental impact and resource consumption.

- Greenhouse Gas Reduction: Lower methane emissions and efficient organic waste utilization contribute to BSF farming's role in mitigating greenhouse gas emissions.

2. Circular Economy Contribution:

- Resource Recovery: BSF larvae convert organic waste into valuable biomass and by-products like fertilizer, supporting a circular economy approach to resource management.

- Economic Opportunities: BSF farming creates opportunities for sustainable business models, waste-to-value chains, and rural development through job creation and economic diversification.

Future Directions and Challenges

Future trends in BSF farming focus on scalability, regulatory frameworks, and consumer acceptance:

1. Scale-Up and Commercialization:

   - Industrial Applications: Scaling BSF farming for industrial applications requires addressing logistical challenges, regulatory compliance, and market acceptance.

   - Global Expansion: Increasing adoption of BSF farming globally hinges on overcoming cultural barriers, regulatory harmonization, and demonstrating economic viability.

2. Research and Innovation:

   - Biotechnological Advances: Continued research explores novel applications of BSF products in biotechnology, pharmaceuticals, and renewable energy sectors.

   - Sustainability Metrics: Developing robust sustainability metrics and lifecycle assessments will validate BSF farming's environmental benefits and support market differentiation.

BSF farming is poised to play a pivotal role in sustainable agriculture, waste management, and innovative bio-based industries. By embracing technological advancements, exploring new applications, and addressing regulatory challenges, BSF farmers can unlock the full potential of this remarkable insect as a catalyst for global sustainability and economic growth. The future of BSF farming holds promise for creating resilient food systems and promoting environmental stewardship in an increasingly resource-constrained world.

Black Soldier Fly (BSF) larvae can contribute to biogas production through their role in organic waste processing. Here's how BSF larvae can be integrated into biogas production processes:

Organic Waste Processing

1. Waste Conversion: BSF larvae are efficient at consuming

organic waste materials such as food scraps, agricultural residues, and manure. They digest these materials, converting them into biomass (larvae) and leaving behind residue (frass).

2. Frass Composition: The residue left after BSF larvae digestion, known as frass, still contains organic matter that can be further processed.

Integration with Biogas Production

3. Anaerobic Digestion: After BSF larvae have processed the organic waste, the remaining frass can be fed into an anaerobic digester. Anaerobic digestion is a biological process where microorganisms break down organic matter in the absence of oxygen, producing biogas as a byproduct.

4. Biogas Composition: Biogas typically consists of methane (CH_4) and carbon dioxide (CO_2), with methane being the primary component. Methane is a potent greenhouse gas but can be captured and used as a renewable energy source.

5. Benefits of BSF Integration:

   - Enhanced Efficiency: BSF larvae can pre-process organic waste, making it more accessible to microorganisms in anaerobic digestion, thus enhancing overall efficiency.

   - Waste Reduction: By converting organic waste into biomass (larvae) and biogas, BSF larvae contribute to waste reduction and environmental sustainability.

   - Nutrient Recycling: The nutrients in the frass can be recycled back into soil as a fertilizer, completing a nutrient cycle that enhances agricultural sustainability.

Practical Applications

6. Scale and Implementation: BSF larvae and anaerobic digestion systems can be implemented at various scales, from small-

scale household digesters to large-scale industrial systems. This flexibility allows for widespread adoption in both urban and rural settings.

7. Environmental Benefits: By diverting organic waste from landfills and reducing methane emissions from waste decomposition, integrating BSF larvae with biogas production supports climate change mitigation efforts.

Conclusion

Integrating BSF larvae into biogas production systems represents a sustainable approach to waste management and renewable energy generation. This integrated approach not only reduces organic waste volumes but also harnesses valuable resources in the form of biomass and biogas. As technology advances and awareness grows, leveraging BSF larvae in biogas production could play a significant role in achieving global sustainability goals and promoting circular economy principles.

Building a biogas-powered plant using Black Soldier Fly (BSF) larvae as a central concept involves integrating two innovative technologies: BSF larvae for organic waste digestion and biogas production through anaerobic digestion. Here's a detailed discussion on how such a plant could be conceptualized and its potential benefits:

Conceptual Framework

1. Organic Waste Collection and Processing:

   - Input Material: The plant would start by collecting organic waste streams suitable for BSF larvae, such as food waste from households, restaurants, and food processing industries, as well as agricultural residues and manure.

   - BSF Larvae Processing: Organic waste is fed to BSF larvae, housed in controlled environments optimized for rapid digestion. The larvae efficiently convert the organic matter into biomass (larvae) and residual frass.

2. Integration with Anaerobic Digestion:

- Frass Utilization: The residual frass, after BSF larvae digestion, is directed into anaerobic digesters. These digesters are sealed, oxygen-free environments where microorganisms break down organic matter in the frass.

- Biogas Production: Anaerobic digestion produces biogas as a byproduct, primarily composed of methane (CH_4) and carbon dioxide (CO_2). Biogas can be captured, stored, and utilized for various purposes, including electricity generation, heating, or as a fuel for vehicles.

Components and Infrastructure

3. Plant Infrastructure:

- BSF Larvae Facilities: The plant would include facilities for BSF larvae rearing, comprising controlled temperature and humidity environments with suitable waste feeding systems.

- Anaerobic Digesters: These would be designed to process the frass efficiently, with monitoring systems to optimize digestion conditions for maximum biogas production.

- Biogas Storage and Utilization: Storage facilities and utilization systems would be in place to store and convert biogas into usable energy, ensuring consistent supply and efficiency.

Operational Considerations

4. Operational Workflow:

- Feedstock Management: Efficient handling and processing of incoming organic waste streams to ensure continuous supply for BSF larvae and anaerobic digestion.

- Monitoring and Control: Implementing monitoring systems to track larvae growth, digestion efficiency, biogas production rates, and overall plant performance.

- Maintenance and Safety: Regular maintenance of equipment and adherence to safety protocols to ensure smooth operations and compliance with environmental regulations.

Benefits and Sustainability

5. Environmental Impact:

- Waste Reduction: By converting organic waste into valuable products (larvae biomass and biogas), the plant contributes to significant waste reduction and landfill diversion.

- Greenhouse Gas Reduction: Anaerobic digestion reduces methane emissions from decomposing organic waste, while biogas utilization displaces fossil fuel use, thus mitigating greenhouse gas emissions.

6. Resource Efficiency:

- Circular Economy: Promotes a circular economy model by recycling nutrients from waste back into agricultural and industrial processes through frass as fertilizer and biogas as energy.

- Water and Land Use: Requires minimal water and land compared to conventional waste management and energy production methods, enhancing resource efficiency.

Economic Viability

7. Revenue Streams:

- Biogas Sales: Selling biogas as a renewable energy source to local industries, utilities, or transportation sectors.

- BSF Products: Generating revenue from BSF larvae biomass as high-protein animal feed supplements or bio-based products.

Building a biogas-powered plant with the concept of integrating BSF larvae digestion offers a sustainable solution to organic waste management and renewable energy production. By leveraging the efficient capabilities of BSF larvae and anaerobic digestion technology, such a plant can contribute significantly to environmental sustainability, resource efficiency, and economic viability. As technologies continue to advance and awareness of sustainable practices grows, investing in such innovative solutions holds promise for addressing global challenges while fostering local economic development and resilience.

Conclusion

Chapter 5 highlights the critical aspects of harvesting, processing, and storing BSF maggots and their meal. By mastering these techniques, farmers and entrepreneurs can ensure product quality, maximize efficiency in production processes, and meet market demands for sustainable protein sources in various industries. Effective harvesting and processing methods not only enhance operational efficiency but also contribute to the economic viability and environmental sustainability of BSF maggot farming ventures.

CHAPTER 6: IMPLEMENTATION AND PRACTICAL APPLICATIONS OF BSF MAGGOT FARMING

In this chapter, we delve into the practical aspects of implementing and applying BSF maggot farming techniques. Building upon the foundational knowledge from earlier chapters, we explore how to establish and manage a BSF maggot farm effectively, as well as various applications and benefits of BSF maggots beyond just protein production.

1. Setting Up Your BSF Maggot Farm

Choosing a Location:

- Evaluate suitable locations for your BSF maggot farm, considering factors such as temperature, humidity, and proximity to organic waste sources.

- Design the farm layout to optimize space utilization and facilitate efficient workflow from egg deposition to larval harvesting.

Infrastructure Requirements:

- Select appropriate containers or racks for rearing BSF larvae, ensuring they are durable, easy to clean, and provide adequate ventilation.

- Install climate control systems if necessary to regulate temperature and humidity, critical for optimal larval development.

Feedstock Management:

- Establish protocols for sourcing and preparing organic waste materials as feedstock for BSF larvae.

- Implement efficient waste collection and segregation practices to ensure consistent and quality feedstock supply.

2. Operational Procedures

Breeding and Hatchery Phase:

- Manage egg deposition and hatchery conditions to maximize egg hatching rates and larval survival.

- Monitor temperature, humidity, and substrate conditions to create an optimal environment for larval growth.

Larval Rearing and Growth:

- Provide a nutrient-rich substrate for larvae to feed on, optimizing growth rates and protein content.

- Adjust feeding schedules and quantities based on larval development stages to maximize biomass production.

Harvesting and Processing:

- Implement harvesting techniques to collect mature larvae efficiently while minimizing contamination.

- Process harvested larvae into dried meal or other forms suitable for further processing or distribution.

3. Applications Beyond Protein Production

Animal Feed Supplement:

- Explore formulations and applications of BSF larvae meal as a high-protein supplement in animal feed for poultry, fish, and livestock.

- Assess nutritional benefits and economic feasibility compared to traditional feed sources.

Biowaste Management:

- Utilize BSF larvae to process organic waste streams, converting waste materials into valuable biomass and reducing environmental impact.

- Evaluate synergies with composting or biogas production systems to enhance overall waste management efficiency.

4. Research and Innovation

Emerging Trends:

- Stay updated on research advancements and innovations in BSF maggot farming, such as genetic improvements or alternative feed sources.

- Collaborate with research institutions or industry partners to explore new applications and sustainable practices.

Market Opportunities:

- Identify niche markets and potential buyers for BSF larvae products, including agricultural industries, pet food manufacturers, and bioindustry sectors.

- Develop marketing strategies to promote the benefits of BSF larvae-based products and differentiate them from conventional alternatives.

5. Challenges and Mitigation Strategies

Regulatory Compliance:

- Navigate regulatory requirements and certifications for insect farming and animal feed production, ensuring adherence to food safety and environmental standards.

- Engage with regulatory authorities and industry associations to address compliance challenges and promote industry best practices.

Operational Efficiency:

- Address challenges related to scalability, efficiency, and cost-effectiveness in BSF maggot farming operations.

- Implement continuous improvement initiatives and technology upgrades to optimize resource utilization and productivity.

Applications of Maggot Meal

Here, we explore the diverse applications of maggot meal, highlighting its versatility as a sustainable protein source in animal feed, its potential for human consumption, and its various industrial applications.

Uses in Animal Feed

1. Poultry Feed:

   - Incorporate maggot meal into poultry diets as a high-protein supplement.

   - Enhance growth rates, feed conversion efficiency, and egg production in chickens.

2. Aquaculture Feed:

   - Substitute fishmeal with maggot meal in aquaculture diets for fish and shrimp.

   - Improve nutritional value and sustainability of fish feed formulations.

3. Livestock Feed:

- Include maggot meal in diets for pigs, cattle, and other livestock to boost protein intake.

- Support healthy growth, milk production, and overall animal health.

Potential for Human Consumption

1. Nutritional Benefits:

   - Explore potential applications of maggot meal as a sustainable protein source for human consumption.

   - Evaluate nutritional composition, including protein content, amino acids, and micronutrients.

2. Culinary Uses:

   - Develop culinary applications and recipes incorporating maggot meal in food products.

   - Address cultural perceptions and consumer acceptance through education and innovative food processing techniques.

Other Industrial Applications

1. Biodegradable Plastics:

   - Research and develop biodegradable plastics using maggot-derived chitin or proteins.

   - Explore potential applications in packaging, biomedical materials, and agricultural films.

2. Organic Fertilizers:

   - Utilize maggot residues and frass (larval excrement) as organic fertilizers.

   - Enhance soil fertility, nutrient availability, and crop yields in sustainable agriculture practices.

3. Waste Management Solutions:

- Integrate maggot farming with waste management systems to process organic waste streams.

- Reduce landfill waste and greenhouse gas emissions through efficient organic waste conversion.

Conclusion

Chapter 6 underscores the multifaceted applications of maggot meal, showcasing its potential in animal feed, human nutrition, and diverse industrial sectors. By exploring innovative uses and advancing research in sustainability and resource efficiency, maggot meal contributes to global efforts in food security, waste management, and environmental conservation. Embracing these applications not only supports economic growth in insect farming but also promotes sustainable development goals through responsible and innovative use of natural resources.

It highlights the practical implementation of BSF maggot farming, from setting up a farm to exploring diverse applications beyond protein production. By mastering operational procedures, leveraging emerging research, and overcoming challenges with innovative solutions, farmers and entrepreneurs can harness the full potential of BSF maggots as a sustainable and versatile resource in agriculture and waste management.

This chapter provides a comprehensive guide to implementing and expanding BSF maggot farming operations, emphasizing practical strategies, applications, and challenges faced in the industry.

CHAPTER 7: SCALING UP AND INNOVATING IN BSF MAGGOT FARMING

In this chapter, we explore strategies for scaling up BSF maggot farming operations and driving innovation in the industry. As interest in sustainable protein sources grows, scaling production while innovating in technology and applications becomes crucial for meeting demand and exploring new opportunities.

1. Scaling Up Your BSF Maggot Farm

Scaling Up Production:

Modular and Expandable Facilities: Design the maggotery in a modular fashion, allowing for gradual expansion as the operation grows. This enables producers to scale up production incrementally and adapt to changing market demands.

Automation and Mechanization: Incorporate automation and mechanization into various aspects of the operation, such as substrate handling, larvae harvesting, and packaging. This can help increase efficiency, reduce labor costs, and improve consistency.

Vertical Integration: Consider integrating various stages of the

BSF value chain, such as waste collection, processing, and product manufacturing. This can create economies of scale and improve the overall profitability of the enterprise.

Expansion Planning:

- Assess market demand and opportunities for scaling up production capacity.

- Develop a phased expansion plan, considering factors like infrastructure, manpower, and financial resources.

Optimizing Production Systems:

- Implement efficient production systems to maximize biomass yield and quality.

- Integrate automation and technology where feasible to streamline operations and reduce labor costs.

Quality Control and Assurance:

- Establish quality control measures for feedstock, larvae development, and final product output.

- Monitor key parameters such as nutritional content, contamination levels, and hygiene standards.

2. Advanced Techniques and Innovations

Genetic Improvement:

- Explore genetic selection and breeding programs to enhance desirable traits in BSF larvae, such as growth rate and nutrient content.

- Collaborate with geneticists or research institutions to leverage biotechnology for genetic improvement.

Innovations in Feeding and Substrate Management:

Diversifying Feedstock: Experiment with a wider range of organic waste streams, including agricultural byproducts, food processing waste, and even specialized formulated diets. This can help optimize the nutritional profile and growth of the larvae.

Substrate Preprocessing: Develop methods to pre-process the organic waste, such as shredding, composting, or fermentation, to enhance the availability of nutrients and improve the larvae's feeding efficiency.

Waste Stream Valorization: Explore ways to extract valuable byproducts from the spent substrate, such as compost or biogas, to generate additional revenue streams and contribute to a more circular economy.

Innovative Rearing Techniques:

Multispecies Systems: Investigate the potential of integrating multiple insect species, such as mealworms or crickets, within the maggotery to diversify production and capitalize on different market demands.

Controlled Environment Systems: Develop more sophisticated environmental control systems, including temperature, humidity, and lighting management, to create optimal conditions for enhanced growth and productivity.

Automated Harvesting and Handling: Implement advanced harvesting and handling technologies, such as robotic systems or automated conveyor belts, to streamline the collection and processing of mature larvae.

Product Development and Value-Addition:

Diversifying BSF-based Products: Explore the development of a broader range of BSF-derived products, such as protein meals, oils, chitin-based materials, or even live larvae for various applications (e.g., pet food, aquaculture, poultry feed).

Branding and Marketing: Establish a strong brand identity for the BSF products, emphasizing their sustainability, nutritional benefits, and environmental impact. Engage in targeted marketing and promotional strategies to reach new market segments.

Partnerships and Collaborations: Seek partnerships with research institutions, technology providers, or industry associations to access the latest innovations, share knowledge, and explore collaborative opportunities for scaling and advancing BSF maggot farming.

By embracing these scaling and innovation strategies, BSF producers can enhance the efficiency, productivity, and profitability of their operations, while contributing to the overall growth and development of the sustainable insect farming industry.

Alternative Feed Sources:

- Investigate alternative feed sources for BSF larvae, including agricultural by-products or food waste blends.

- Conduct trials to optimize nutrient composition and larval performance with different feed formulations.

Integrated Systems Approach:

- Integrate BSF maggot farming with other agricultural or waste management systems, such as composting or biogas production.

- Explore synergies to enhance resource efficiency and sustainability across multiple sectors.

3. Exploring New Applications

Bioconversion Technologies:

- Partner with biotechnology firms or waste management companies to explore bioconversion technologies using BSF larvae.

- Investigate the potential for extracting valuable compounds or producing bio-based materials from BSF larvae biomass.

Aquaculture and Livestock Feed:

- Expand applications in aquaculture and livestock industries by promoting BSF larvae meal as a sustainable protein source.

- Collaborate with feed manufacturers to develop tailored formulations and market BSF larvae products effectively.

4. Market Development and Commercialization

Market Analysis:

- Conduct market research to identify niche markets, trends, and consumer preferences for insect-based products.

- Develop targeted marketing strategies to educate and engage potential buyers in agriculture, pet food, and bioindustry sectors.

Certifications and Regulations:

- Navigate regulatory frameworks for insect farming and animal feed production, ensuring compliance with food safety and environmental standards.

- Pursue certifications and endorsements to build credibility and access international markets.

5. Sustainability and Corporate Responsibility

Environmental Impact Assessment:

- Assess the environmental footprint of BSF maggot farming operations, including waste reduction and greenhouse gas emissions.

- Implement sustainable practices to minimize ecological impact and promote biodiversity conservation.

Corporate Social Responsibility:

- Engage with local communities and stakeholders to promote awareness of sustainable agriculture and insect-based nutrition.

- Support initiatives that contribute to social welfare, education, and economic development in farming communities.

The economics of maggot farming

The economics of maggot farming involves analyzing the financial aspects of cultivating and harvesting maggots for various applications. Maggots are increasingly being used in industries such as agriculture, waste management, and animal feed production due to their ability to efficiently convert organic waste into high-protein biomass.

When considering the economics of maggot farming, factors such as initial start-up costs, operating expenses (such as feeding and maintaining the maggot colonies), revenue streams (such as the sale of live maggots or processed products), and market demand need to be taken into account.

The profitability of maggot farming can vary depending on various factors such as the scale of the operation, production efficiency, market prices for maggots and related products, as well as regulatory requirements and competition within the industry.

Overall, the economics of maggot farming can be financially viable for businesses looking to utilize maggots as a sustainable and cost-effective solution for waste management or as a source of high-quality protein for animal feed. However, careful financial planning and market analysis are essential to ensure the success of a maggot farming venture.

Economic and Environmental Benefits

We will examine the economic viability, environmental impact, and sustainability of maggot farming. We explore cost analysis and profitability metrics, environmental benefits, and showcase case studies of successful maggot farms around the world.

Cost Analysis and Profitability

1. Production Costs:

   - Evaluate initial investment costs, including infrastructure, equipment, and labor for setting up a maggot farm.

   - Analyze ongoing operational expenses such as feedstock procurement, utilities, and maintenance.

2. Revenue Streams:

   - Calculate potential revenue from sales of maggot meal, larvae, or other by-products to various markets.

   - Explore pricing strategies and market demand dynamics influencing profitability.

3. Profitability Metrics:

   - Conduct financial analysis to assess return on investment (ROI) and profitability margins.

   - Compare cost-efficiency of maggot farming with traditional livestock or protein production systems.

Environmental Impact and Sustainability

1. Waste Reduction and Recycling:

   - Highlight environmental benefits of maggot farming in converting organic waste into valuable protein and nutrient-rich biomass.

   - Quantify reduction in landfill waste and greenhouse gas emissions compared to conventional waste disposal methods.

2. Resource Efficiency:

   - Discuss efficiency in resource utilization, including land, water, and feedstock inputs, compared to conventional animal farming.

   - Measure carbon footprint and environmental footprint metrics associated with maggot farming operations.

3. Biodiversity and Ecosystem Services:

   - Explore positive impacts on biodiversity conservation and ecosystem services through sustainable agricultural practices.

   - Showcase examples of habitat restoration and ecological benefits observed in maggot farming environments.

Case Studies of Successful Maggot Farms

1. Global Perspectives:

   - Present case studies of pioneering maggot farms across different continents and climates.

   - Highlight success stories, challenges faced, and lessons learned in scaling up operations and market penetration.

2. Local Innovations:

   - Feature innovative practices and technologies adopted by maggot farms in specific regions or countries.

   - Illustrate adaptation to local market conditions, regulatory frameworks, and community engagement strategies.

3. Future Outlook:

   - Discuss emerging trends and future prospects for maggot farming in sustainable agriculture and food production.

   - Predict potential growth areas, market expansion opportunities, and technological advancements shaping the industry.

AgriProtein (South Africa):

AgriProtein is a leading company in the insect-based feed industry, specializing in large-scale maggot farming.

The company has successfully built commercial-scale facilities that can process up to 250 tons of organic waste per day, converting it into high-protein animal feed.

AgriProtein's products, such as MagMeal and MagOil, are widely recognized for their quality and have gained traction in the global animal feed market.

The company has expanded its operations to several countries, including the United States, Australia, and the Netherlands, demonstrating the scalability and replicability of their maggot farming model.

Entomo Farms (Canada):

Entomo Farms is a Canadian company that has pioneered the commercial-scale production of edible insects, including maggots.

The farm utilizes a modular and automated system to efficiently rear black soldier fly larvae, processing over 36 tons of organic waste per day.

Entomo Farms' maggot-based products, such as protein powders and whole dried larvae, are sold to the pet food, aquaculture, and human food industries.

The company's focus on innovation and sustainability has earned it recognition, including the Cleantech Open Global Forum Award and the Deloitte Technology Fast 50 award.

Musca Domestica (Indonesia):

Musca Domestica is an Indonesian startup that has successfully scaled up its maggot farming operations to process over 5 tons of market waste per day.

The company's integrated approach includes collecting organic waste from local markets, rearing the black soldier fly larvae, and producing high-protein animal feed.

Musca Domestica's maggot-based feed has gained popularity among local poultry and fish farmers, who have reported improved animal health and productivity.

The startup's model demonstrates the potential for maggot

farming to address waste management challenges and create economic opportunities in developing regions.

Hexafly (Ireland):

Hexafly is an Irish company that has developed a proprietary technology for the efficient production of black soldier fly larvae.

The company's modular and automated systems enable the processing of up to 50 tons of organic waste per day, converting it into protein-rich animal feed and biofertilizer.

Hexafly's products have gained recognition for their sustainability, with the company receiving numerous awards, including the Irish Times Innovation Award and the Enterprise Ireland Innovation Award.

The company's focus on research and development has led to the optimization of various aspects of maggot farming, including substrate management and environmental control.

These case studies demonstrate the potential for maggot farming to become a scalable and commercially viable solution for waste management, animal feed production, and the creation of a more circular economy.

Chapter 7 emphasizes the importance of scaling up BSF maggot farming operations while fostering innovation in technology, applications, and market development. By adopting advanced techniques, exploring new applications, and maintaining a commitment to sustainability, stakeholders in the BSF farming industry can drive growth, meet evolving market demands, and contribute to global food security and environmental sustainability.

This chapter provides insights into scaling up BSF maggot farming operations and exploring innovative pathways to expand market reach and impact in sustainable agriculture and waste management sectors.

Chapter 7 underscores the economic resilience, environmental stewardship, and innovation potential of maggot farming. By showcasing cost-effectiveness, environmental benefits, and real-world success stories, maggot farming emerges as a transformative solution in sustainable agriculture, waste management, and protein production. Embracing these benefits fosters a resilient and sustainable food system while promoting global efforts in resource efficiency and environmental conservation.

CHAPTER 8: CHALLENGES AND FUTURE DIRECTIONS IN MAGGOT FARMING

In this chapter, we explore the challenges currently facing maggot farming and look towards future directions for the industry. As maggot farming gains traction as a sustainable protein source and waste management solution, addressing these challenges and seizing opportunities for innovation are crucial for its continued growth and impact.

Current Challenges in Maggot Farming

1. Regulatory Frameworks:

   - Navigate complex regulatory environments governing insect farming and feed production.

   - Address regulatory barriers and inconsistencies across different regions or countries.

2. Consumer Perception:

   - Overcome cultural perceptions and consumer acceptance of insect-based products, including maggot meal.

   - Educate consumers about the nutritional benefits and sustainability of maggot-derived products.

3. Scaling Production:

- Manage challenges related to scaling up production capacity while maintaining product quality and consistency.

- Address logistical constraints, including feedstock availability, labor, and infrastructure requirements.

4. Technological Innovation:

- Invest in research and development of advanced technologies for optimizing larval growth, feed conversion efficiency, and waste processing.

- Explore automation and digital solutions to enhance operational efficiency and reduce production costs.

Future Directions and Innovations

1. Genetic Enhancement:

- Explore genetic selection and breeding programs to improve desirable traits in BSF larvae, such as growth rate, nutrient content, and resistance to environmental stressors.

- Collaborate with biotechnologists and geneticists to accelerate genetic improvement in maggot farming.

2. Novel Applications:

- Expand applications of maggot-derived products beyond animal feed to include pharmaceuticals, nutraceuticals, and bioplastics.

- Explore synergies with circular economy principles and sustainable product development.

3. Integrated Systems Approach:

- Integrate maggot farming with other sustainable agriculture practices, such as organic farming, aquaculture, and biogas production.

- Develop integrated farming models that optimize resource use and enhance ecological resilience.

4. Global Collaboration and Knowledge Sharing:

   - Foster international collaboration among researchers, industry stakeholders, and policymakers to share best practices and innovations in maggot farming.

   - Support capacity-building initiatives and knowledge exchange programs to promote sustainable development goals.

Challenges in Maggot Farming:

1. Regulation and Certification: The lack of standardized regulations and certification processes for maggot farming can pose challenges for producers in terms of meeting quality and safety standards.

2. Market Acceptance: Convincing consumers and industries to adopt maggot-based products can be a hurdle due to perceived social stigmas or lack of awareness about the benefits of maggot farming.

3. Scaling Up Production: Scaling up maggot farming operations can be challenging as it requires significant investment in infrastructure, technologies, and skilled labor.

4. Feed Availability: Maggots require specific feed sources, and ensuring a consistent and sustainable supply of feed can be a challenge, especially in regions with limited resources.

5. Competition: As the maggot farming industry grows, competition among producers may increase, leading to price pressures and a need for differentiation strategies.

Future Directions in Maggot Farming:

1. Research and Development: Continued research into optimizing maggot production processes, exploring new feed sources, and developing value-added products can drive innovation and growth in the industry.

2. Sustainability: Implementing sustainable practices in maggot farming, such as utilizing organic waste as feed and minimizing environmental impacts, will be crucial for long-term success.

3. Market Diversification: Exploring new markets and applications for maggot-derived products, such as pharmaceuticals, cosmetics, and biofuels, can open up new opportunities for revenue growth.

4. Technology Adoption: Leveraging advancements in technologies such as automation, data analytics, and bioengineering can enhance efficiency and productivity in maggot farming operations.

5. Education and Outreach: Increasing awareness and education about the benefits of maggot farming for sustainability and waste management can help in overcoming social barriers and expanding market acceptance.

Startups and companies are exploring modular and automated systems to streamline production and overcome scalability hurdles.

Regulatory Hurdles:

The use of insects in animal feed and food products is subject to strict regulations in many regions, which can slow the adoption and commercialization of maggot farming.

Regulatory bodies are gradually updating their policies to recognize the potential of insect-based products, but navigating the regulatory landscape remains a key challenge.

Substrate Optimization:

Ensuring a consistent and reliable supply of suitable organic waste substrates is crucial for the success of maggot farms.

Optimizing substrate composition, processing, and storage can be complex, as it affects the growth and nutritional value of the maggots.

Process Efficiency:

Improving the efficiency of maggot farming processes, such as larval rearing, harvesting, and post-processing, is essential to enhance profitability and sustainability.

Innovations in automation, temperature and humidity control, and waste management can contribute to increased process efficiency.

Market Acceptance:

Overcoming the stigma and misconceptions surrounding the use of insects in animal feed and human food is a significant challenge.

Extensive educational campaigns and collaboration with industry stakeholders are needed to promote the benefits of maggot-based products and increase consumer acceptance.

Future Directions:

Diversification of Applications:

Exploring the use of maggot-derived products beyond animal feed, such as in the cosmetics, pharmaceutical, and biofuel industries, can expand the market opportunities for maggot farming.

Genetic Improvements:

Advancements in genetic engineering and selective breeding techniques can lead to the development of improved black soldier fly strains with enhanced growth rates, nutrient profiles, and waste-conversion capabilities.

Integrated Waste Management Systems:

Integrating maggot farming into comprehensive waste

management systems, where various organic waste streams are utilized for multiple purposes, can maximize resource efficiency and sustainability.

Collaboration and Knowledge Sharing:

Fostering collaboration between researchers, entrepreneurs, and policymakers can accelerate the development and adoption of maggot farming technologies.

Sharing best practices, research findings, and innovative approaches can help the industry overcome challenges and shape a more sustainable future.

Regulatory Harmonization:

Streamlining and harmonizing regulations related to insect-based products across different regions can facilitate the global expansion of maggot farming and its associated industries.

By addressing these challenges and pursuing the outlined future directions, the maggot farming industry can unlock its full potential, contributing to a more circular and sustainable agricultural ecosystem.

Chapter 8 highlights the ongoing challenges and promising future directions in maggot farming. By addressing regulatory, technological, and market challenges while embracing innovation and collaboration, the industry can unlock its full potential as a sustainable solution in food production, waste management, and environmental conservation. As global demand for protein and sustainable practices continues to grow, maggot farming stands poised to play a pivotal role in shaping the future of agriculture and fostering resilience in the face of global challenges.

CHAPTER 9: THE PATH FORWARD: ADVANCING MAGGOT FARMING FOR SUSTAINABLE DEVELOPMENT

In this chapter, we explore the strategic pathways and actionable steps needed to advance maggot farming as a cornerstone of sustainable development. Building upon the insights and challenges discussed earlier, Chapter 9 focuses on fostering innovation, promoting adoption, and maximizing the positive impact of maggot farming on global food security, environmental sustainability, and economic growth.

1. Harnessing Innovation in Maggot Farming

1. Research and Development:

Enhancing BSF Biology and Breeding:

Conduct in-depth studies on the life cycle, genetics, and

nutritional requirements of black soldier flies (BSF) to optimize breeding and rearing protocols.

Explore genetic engineering and selective breeding techniques to develop BSF strains with improved growth rates, feed conversion efficiency, and nutrient profiles.

Investigate the influence of environmental factors, such as temperature, humidity, and light, on BSF development and productivity.

2. Technological Integration:

Design automated, modular, and scalable farming systems to streamline maggot production processes, from waste collection to harvesting and processing.

Leverage IoT sensors, data analytics, and AI-driven decision-making to monitor and optimize farm operations, resource utilization, and waste management.

Explore novel techniques for substrate pre-treatment, larval separation, and post-processing to enhance efficiency and product quality.

Diversify maggot-derived products beyond animal feed, such as biofuels, cosmetics, and pharmaceutical applications.

2. Promoting Market Adoption and Consumer Acceptance

Promoting Market Adoption and Consumer Acceptance:

2.1 Targeted Marketing and Education:

Develop engaging marketing campaigns that highlight the nutritional, environmental, and economic benefits of maggot-based products.

Collaborate with chefs, food scientists, and social media influencers to create innovative culinary experiences and recipes featuring maggot meal.

Organize educational workshops, public events, and interactive media content to demystify and destigmatize the use of insects in food and feed.

2.2 Regulatory Advocacy:

Work closely with policymakers, regulatory bodies, and industry associations to advocate for favorable policies and regulations that support the growth of the maggot farming industry.

Participate in the development of international standards, certifications, and guidelines for the production, processing, and use of insect-based products.

Engage in public-private partnerships to navigate complex regulatory landscapes and facilitate market access for maggot-derived products.

Scaling Up Production and Industry Collaboration:

3.1 Strategic Infrastructure Development:

Invest in large-scale, state-of-the-art maggot farming facilities equipped with advanced automation, waste management systems, and processing capabilities.

Establish regional hubs and decentralized production networks to enhance supply chain resilience and reduce transportation costs.

Foster international collaborations and joint ventures to leverage expertise, share best practices, and accelerate the global expansion of the maggot farming industry.

3.2 Supply Chain Optimization:

Diversify feedstock sources and establish reliable supply agreements with organic waste producers to ensure a consistent and sustainable supply of substrates.

Implement blockchain-based traceability systems to enhance product transparency, quality assurance, and regulatory compliance.

Develop industry-wide certification schemes and quality standards to build trust and confidence among consumers and stakeholders.

Sustainability and Environmental Stewardship:

4.1 Circular Economy Integration:

Integrate maggot farming into comprehensive circular economy models, where organic waste streams are utilized as feedstock and maggot-derived products are reintegrated into various industries.

Explore innovative ways to convert maggot biomass into high-value bioproducts, such as biofuels, bioplastics, and soil amendments.

Promote synergies between maggot farming and other agricultural or industrial sectors to maximize resource efficiency and minimize waste.

By harnessing innovation across these key areas, the maggot farming industry can unlock its full potential, driving sustainable growth, environmental stewardship, and the creation of a more circular and resilient food and agricultural system. - Integrate maggot farming with circular economy principles by utilizing organic waste streams as feedstock and generating value-added products.

   - Promote closed-loop systems that minimize waste generation and maximize resource efficiency in agricultural and industrial sectors.

4.2 Climate Resilience:

   - Mitigate climate change impacts by promoting climate-smart farming practices and reducing greenhouse gas emissions associated with conventional agriculture.

- Monitor and adapt to climate variability to ensure long-term sustainability and resilience of maggot farming operations.

Conclusion: Shaping a Sustainable Future with Maggot Farming

Chapter 9 underscores the transformative potential of maggot farming in addressing global challenges related to food security, environmental sustainability, and economic development. By embracing innovation, fostering market adoption, and promoting sustainable practices, stakeholders can collectively advance maggot farming as a resilient and scalable solution in the global agricultural landscape. Together, we can build a future where maggot farming not only meets the nutritional needs of a growing population but also contributes to building resilient food systems and preserving our planet's natural resources for future generations.

CHAPTER 10: THE ROLE OF INDIVIDUALS AND COMMUNITIES IN PROMOTING MAGGOT FARMING

In this chapter, we explore the pivotal role of individuals, communities, and small-scale enterprises in driving the adoption and success of maggot farming. From backyard enthusiasts to community cooperatives and entrepreneurial ventures, Chapter 10 highlights the transformative power of grassroots initiatives in advancing sustainable agriculture and food security.

1. Empowering Individuals as Change Agents

1.1 Backyard Farming Initiatives:

- Encourage individuals to start small-scale maggot farms in their backyard or urban settings.

- Provide guidance on setup, management, and utilization of

locally available resources.

1.2 Education and Awareness:

- Promote education programs and workshops to inform communities about the benefits and methods of maggot farming.

- Empower individuals with practical knowledge to integrate maggot farming into sustainable living practices.

2. Community-Based Enterprises and Cooperatives

2.1 Collaborative Farming Models:

- Facilitate the formation of community-based cooperatives or enterprises focused on maggot farming.

- Foster collaboration in resource sharing, collective purchasing power, and market access.

2.2 Socio-Economic Benefits:

- Highlight socio-economic benefits of community-led maggot farming initiatives, including job creation, income generation, and local economic development.

- Empower marginalized communities and women entrepreneurs through inclusive participation in maggot farming ventures.

3. Supporting Entrepreneurship and Innovation

3.1 Incubation and Support Programs:

- Establish incubation centers and support programs for aspiring entrepreneurs in the insect farming sector.

- Provide access to financing, mentorship, and technical

assistance to scale innovative maggot farming ventures.

3.2 Innovation Hubs and Research Collaboration:

- Foster innovation hubs where researchers, startups, and industry leaders collaborate on advancing maggot farming technologies and applications.

- Encourage open innovation and knowledge exchange to accelerate industry growth and market innovation.

4. Advocacy for Policy and Regulatory Support

4.1 Policy Advocacy:

- Advocate for policies that promote the adoption of insect farming, including maggot farming, as a sustainable agricultural practice.

- Engage policymakers in dialogue to address regulatory barriers and create an enabling environment for industry growth.

4.2 Standards and Certification:

- Collaborate with industry associations and regulatory bodies to establish standards and certifications for maggot-derived products.

- Ensure compliance with food safety, environmental sustainability, and ethical practices to build consumer trust and market acceptance.

Conclusion: Building a Sustainable Future Together

Chapter 10 underscores the transformative potential of grassroots initiatives and community-driven efforts in promoting maggot farming as a sustainable solution for food security

and environmental stewardship. By empowering individuals, supporting community enterprises, fostering entrepreneurship, and advocating for supportive policies, we can collectively harness the benefits of maggot farming to create resilient food systems and a sustainable future for generations to come. Together, let us embrace the opportunities of maggot farming to nourish our communities, protect our planet, and shape a brighter tomorrow
.

CHAPTER 11: LOOKING AHEAD: FUTURE TRENDS AND OPPORTUNITIES IN MAGGOT FARMING

In this chapter, we explore the emerging trends, future opportunities, and potential challenges that will shape the landscape of maggot farming in the coming years. As the world seeks sustainable solutions to global challenges, maggot farming stands poised to play a pivotal role in food security, waste management, and economic development.

1. Technological Advancements and Innovation

1.1 Automation and Digitalization:

 - Embrace advancements in automation and digital technologies to optimize farm management, feeding regimes, and environmental controls.

 - Implement IoT (Internet of Things) and AI (Artificial

Intelligence) solutions for real-time monitoring, predictive analytics, and decision-making.

1.2 Biotechnological Innovations:

- Explore biotechnological advancements for enhancing nutrient composition, protein content, and bioactive compounds in maggot-derived products.

- Leverage biorefinery approaches to extract valuable bio-based materials from maggot biomass for pharmaceuticals, nutraceuticals, and bioplastics.

2. Market Expansion and Diversification

2.1 Global Market Growth:

- Tap into expanding global markets for insect-based products, including animal feed, human nutrition, and industrial applications.

- Develop export strategies and partnerships to access new markets and capitalize on increasing consumer interest in sustainable protein sources.

2.2 Consumer Acceptance and Education:

- Continue efforts to educate consumers about the nutritional benefits, sustainability, and safety of maggot-derived products.

- Collaborate with chefs, influencers, and food innovators to create appealing culinary experiences and increase consumer acceptance.

3. Sustainable Development and Environmental Impact

3.1 Circular Economy Integration:

- Strengthen integration of maggot farming with circular economy principles by utilizing organic waste streams and promoting resource efficiency.

- Explore opportunities for carbon sequestration, nutrient cycling, and ecosystem services in maggot farming systems.

3.2 Climate Resilience and Adaptation:

- Mitigate climate change impacts through resilient farming practices and adaptation strategies tailored to local environmental conditions.

- Foster climate-smart agriculture initiatives that enhance adaptive capacity and sustainability of maggot farming operations.

4. Policy Support and Regulatory Frameworks

4.1 Advocacy and Policy Development:

- Advocate for supportive policies, incentives, and regulatory frameworks that recognize the role of maggot farming in sustainable agriculture.

- Engage policymakers, stakeholders, and industry leaders to address barriers and promote a favorable operating environment for insect farming.

Conclusion: Embracing the Future of Maggot Farming

Chapter 11 underscores the transformative potential of maggot farming in addressing global challenges and advancing sustainable development goals. By embracing technological innovation, expanding market opportunities, promoting environmental stewardship, and advocating for supportive policies, stakeholders can collectively shape a resilient and prosperous future for maggot farming. Together, let us seize the opportunities ahead to harness the full potential of maggot farming as a sustainable, scalable, and impactful solution in the global agricultural landscape.

CHAPTER 12: CONCLUSION: EMBRACING THE POTENTIAL OF MAGGOT FARMING

In this final chapter, we reflect on the journey through the world of maggot farming, exploring its transformative potential, benefits, challenges, and future outlook. Chapter 12 serves as a culmination of our exploration, highlighting the key takeaways and urging stakeholders to embrace maggot farming as a sustainable solution for the future.

Recapitulating the Journey

Maggot farming, also known as "maggotry", is the process of raising fly larvae (maggots) for various purposes, including animal feed, waste management, and bioremediation.

The most commonly used fly species for maggot farming are the house fly (Musca domestica) and the black soldier fly (Hermetia illucens). These species are preferred due to their rapid growth, ability to thrive on a variety of organic waste substrates, and nutritional value.

Maggot farming can be a profitable business venture, particularly in regions like Rivers State, South-South Nigeria, where it has been successfully implemented as a sustainable commercial enterprise for producing animal and aquafeed.

Key factors for successful maggot farming include the selection of appropriate substrates (e.g., animal manures, agricultural waste, food waste), maintaining optimal environmental conditions (temperature, humidity, aeration), and efficient harvesting and processing of the maggots.

Maggots produced through this process can be used as a high-protein, nutrient-rich feed for livestock, poultry, and aquaculture, providing a sustainable and cost-effective alternative to conventional feed sources.

Maggot farming also contributes to waste management by converting organic waste into valuable biomass, reducing the environmental impact of waste disposal.

Research has been conducted on factors influencing the sex ratio and reproductive biology of the black soldier fly, which is important for optimizing maggot production.

The available literature provides comprehensive information on the technical aspects, potential benefits, and practical implementation of sustainable commercial maggot production for various applications.

From Waste to Resource:

   - Maggot farming transforms organic waste into valuable protein and nutrient-rich biomass.

   - It exemplifies circular economy principles by closing the nutrient loop and reducing environmental impact.

This topic aligns well with the principles of the circular economy and sustainable waste management. The core idea behind this approach is to view organic waste not as a liability or burden, but as a resource with inherent potential. Maggot farming provides an effective and efficient way to harness that potential and convert waste into useful, high-value products.

One of the primary benefits of maggot farming is its ability to effectively manage and process various organic waste streams, such as agricultural waste, food waste, and animal manure. These waste materials, which would otherwise end up in landfills or causing environmental degradation, can be transformed into a protein-rich biomass through the feeding and growth of maggots.

This process of nutrient recycling is particularly valuable, as the maggot biomass can then be utilized as a high-quality animal feed or aquafeed, providing a sustainable and cost-effective alternative to conventional feed sources. By reintroducing these nutrients back into the food production system, maggot farming creates a circular economy, where waste is not discarded but rather repurposed and reintegrated into the supply chain.

Beyond the direct use of maggot biomass as feed, the process of maggot farming can also contribute to bioremediation efforts. Certain maggot species, like the black soldier fly, have the ability to degrade and transform complex organic compounds, making them useful for the treatment of wastewater or contaminated soils. This process not only reduces the environmental impact of waste but also helps to restore and regenerate damaged ecosystems.

Furthermore, the volume reduction achieved through maggot farming is a significant benefit. As maggots convert a large portion of the organic waste into their own body mass, the overall volume of the waste stream is dramatically reduced, easing the burden on waste management systems and potentially reducing disposal costs.

From an economic perspective, the transformation of waste

into valuable resources can create new income streams and business opportunities, making maggot farming a sustainable and profitable enterprise. This aligns with the broader goals of the circular economy, which seeks to maximize the value of resources and minimize waste throughout the entire life cycle.

To fully realize the potential of this approach, it is essential to scale up maggot farming operations, moving from small-scale efforts to larger commercial facilities. This will allow for the efficient processing of larger volumes of organic waste and the production of significant quantities of maggot-based products, further enhancing the environmental and economic benefits.

In conclusion, the concept of transforming waste into a valuable resource through maggot farming is a compelling and impactful approach to sustainable waste management. By harnessing the natural capabilities of maggots, we can create a circular economy that promotes nutrient recycling, bioremediation, and the efficient utilization of organic waste streams. This holistic perspective aligns with the principles of sustainability and the transition towards a more circular and resource-efficient future.

Efficiency and Sustainability:

   - Maggots are efficient converters of feedstock, requiring less land, water, and resources compared to traditional livestock.

   - They contribute to greenhouse gas reduction and promote biodiversity conservation through sustainable practices.

Efficiency:

One of the key drivers behind the success of maggot farming as a waste-to-resource approach is its inherent efficiency. Maggots are remarkably adept at converting organic waste into their own biomass, which can then be harvested and utilized. This conversion process is highly efficient, with maggots capable of converting up to 50% of the waste they consume into their own body weight.

This efficiency translates to several tangible benefits:

Reduced waste volumes: The maggots' ability to rapidly convert waste into their own biomass results in a significant reduction in the overall volume of the waste stream, easing the burden on waste management systems.

Faster processing times: Maggot farming can process organic waste much more rapidly compared to traditional composting or other decomposition methods, allowing for quicker transformation and utilization of the waste.

Scalability: The efficient nature of maggot farming makes it highly scalable, enabling the processing of large volumes of waste at commercial-scale facilities. This scalability is crucial for realizing the full potential of this approach.

Sustainability:

Beyond efficiency, the sustainability aspects of maggot farming are equally compelling. This approach aligns with the principles of the circular economy, which emphasizes the need to keep resources in use for as long as possible, extract the maximum value from them, and then recover and regenerate products and materials at the end of their service life.

The key sustainability features of maggot farming include:

Nutrient recycling: By converting organic waste into a nutrient-rich biomass, maggot farming enables the recycling of valuable nutrients back into the food production system, reducing the need for external inputs and promoting a more sustainable agricultural model.

Waste reduction: Diverting organic waste from landfills or open dumping and transforming it into useful products helps to minimize the environmental impact of waste, reducing greenhouse gas emissions and conserving natural resources.

Bioremediation: The ability of certain maggot species to degrade and transform complex organic compounds makes them valuable for bioremediation applications, which can help restore and regenerate damaged ecosystems.

Economic viability: The transformation of waste into valuable resources creates new income streams and business opportunities, making maggot farming a sustainable and profitable enterprise.

Ultimately, the efficiency and sustainability of maggot farming are inextricably linked. The inherent efficiency of the process enables the scaling and widespread adoption of this approach, while the sustainability features ensure that it aligns with broader environmental and economic goals. By harnessing these synergies, maggot farming can play a crucial role in the transition towards a more circular and resource-efficient future.

2. Economic Opportunities and Market Potential

2.1 Diverse Applications:

- Maggot meal serves as a sustainable protein source in animal feed, potentially supplementing or replacing conventional feed ingredients.

- There is growing interest in human consumption applications, supported by nutritional benefits and culinary innovation.

2.2 Market Expansion:

- The global market for insect-based products, including maggot-derived items, is poised for growth amid increasing demand for sustainable protein sources.

- Opportunities exist for entrepreneurship, job creation, and economic development in maggot farming ventures worldwide.

This is a critical consideration as the industry seeks to scale up and maximize the impact of this innovative approach.

Market expansion in the context of maggot farming can take several forms:

Diversification of waste streams:

Currently, maggot farming is primarily focused on processing organic waste streams like agricultural waste, food waste, and animal manure.

However, there is significant potential to expand the range of waste streams that can be effectively converted into valuable resources.

Exploring the use of maggots for the processing of other types of organic waste, such as municipal solid waste, sewage sludge, or even certain industrial waste streams, could significantly broaden the market opportunities.

This diversification would allow maggot farming to address a wider range of waste management challenges and cater to a broader customer base.

Product diversification:

While the primary focus has been on the production of maggot-based animal feed and aquafeed, there are opportunities to explore additional high-value products derived from maggot farming.

Potential products could include biofuels, bioplastics, fertilizers, or even nutraceuticals and pharmaceuticals derived from the unique properties of maggots.

Diversifying the product portfolio would enable maggot farming enterprises to expand their revenue streams and cater to a wider range of industries and applications.

Geographic expansion:

Maggot farming is currently more prevalent in certain regions, such as parts of Asia and Africa, where waste management challenges and the demand for alternative protein sources are particularly acute.

However, there is significant potential for the expansion of maggot farming into new geographic markets, particularly in developed economies where waste management and sustainability concerns are gaining increasing attention.

Adapting the technology and business models to local contexts, addressing regulatory frameworks, and building partnerships with local stakeholders would be crucial for successful geographic expansion.

Scalability and industrialization:

As mentioned earlier, the inherent efficiency of maggot farming makes it highly scalable, allowing for the processing of large volumes of waste at commercial-scale facilities.

Investing in the development of larger, more automated maggot farming operations could unlock significant economies of scale, leading to cost reductions and increased profitability.

This industrialization of the maggot farming industry would enable greater market penetration and the ability to serve larger, more diverse customer bases.

By embracing these market expansion strategies, the maggot farming industry can unlock significant growth opportunities and solidify its position as a key player in the transition towards a more circular and sustainable waste management ecosystem. This would not only benefit the industry itself but also contribute to broader environmental and economic goals on a global scale.

3. Overcoming Challenges and Building Resilience

the challenges and the need to build resilience in the maggot farming industry, particularly around the regulatory landscape and consumer perception. Let's delve deeper into these critical aspects:

Regulatory Landscape:

Addressing regulatory complexities and promoting supportive policies are indeed crucial for the growth and market acceptance of maggot farming.

The regulatory environment can vary significantly across different regions and jurisdictions, presenting a complex landscape for maggot farming enterprises to navigate.

Key regulatory considerations may include:

Waste management regulations: Ensuring compliance with laws and guidelines related to the handling, processing, and disposal of organic waste streams.

Food and feed safety standards: Adhering to regulations and certifications for the use of maggot-derived products in animal feed, human food, or other applications.

Environmental regulations: Meeting requirements related to emissions, water usage, and overall environmental impact.

Occupational health and safety: Addressing worker safety and hygiene protocols within maggot farming facilities.

Collaboration with policymakers, research institutions, and industry stakeholders is essential for shaping supportive regulatory frameworks that foster the growth of the maggot farming industry.

This may involve actively engaging with regulatory bodies, providing scientific evidence, and advocating for policy reforms that recognize the benefits and sustainability of maggot farming.

Consumer Perception:

Addressing and improving consumer perception is another critical challenge that the maggot farming industry must overcome.

Historically, the use of insects, including maggots, in food and feed has faced some degree of cultural stigma and resistance from consumers.

Overcoming this perception will require a multi-pronged approach:

Education and awareness campaigns: Informing consumers about the nutritional, environmental, and economic benefits of maggot-derived products.

Transparency and communication: Fostering trust through open and transparent communication about the maggot farming process, quality control, and safety measures.

Product branding and positioning: Developing appealing and innovative product presentations that destigmatize the use of maggots and highlight their positive attributes.

Collaboration with influential stakeholders: Partnering with respected organizations, chefs, or public figures to endorse and promote the acceptance of maggot-based products.

Building consumer confidence and acceptance will be crucial for the widespread adoption and market success of maggot farming.

Building Resilience:

Addressing these regulatory and consumer perception challenges will be critical for building resilience within the maggot farming industry.

Resilience can be further strengthened through:

Diversification of waste streams and product portfolios: Reducing reliance on a single waste source or product line and expanding into new markets.

Technological innovation: Investing in research and development to improve efficiency, scalability, and product quality.

Supply chain integration: Establishing robust and reliable supply chains to ensure consistent access to raw materials and distribution channels.

Stakeholder collaboration: Fostering partnerships with research institutions, policymakers, and industry players to collectively address shared challenges.

By proactively addressing the regulatory landscape and consumer perception, while simultaneously building resilience through diversification, innovation, and collaboration, the maggot farming industry can position itself for long-term growth, sustainability, and market dominance. 4. Innovating for the Future

4.1 Technological Advancements:

- Embracing technological innovations such as automation, digitalization, and biotechnological enhancements can optimize production efficiency and product quality.

- Research and development efforts should focus on improving larval performance, nutrient profiles, and exploring new applications.

4.2 Sustainability Commitment:

- Upholding environmental stewardship through sustainable farming practices, waste management solutions, and climate resilience initiatives remains paramount.

- Integrating maggot farming into broader sustainability agendas can enhance its role in mitigating climate change and promoting sustainable food systems.

Conclusion: Embracing the Potential

Chapter 12 concludes with a call to action for stakeholders across sectors to embrace the potential of maggot farming. By fostering innovation, promoting market adoption, advocating for supportive policies, and prioritizing sustainability, we can collectively harness the transformative power of maggot farming to address global challenges and build a resilient future. Together, let us continue to innovate, collaborate, and advocate for a sustainable and prosperous world empowered by maggot farming.

GLOSSARY

1. Black Soldier Fly (BSF):
   - Hermetia illucens, a species valued for converting organic waste into protein-rich larvae.

2. Larvae:
   - The immature stage of insects, such as BSF larvae, used for waste decomposition and protein production.

3. Maggot Meal:
   - Dried and processed larvae used as a protein-rich feed ingredient in animal diets.

4. Frass:
   - Excrement or waste produced by larvae during feeding, rich in nutrients and used as fertilizer.

5. Bioconversion:
   - Process of converting organic materials into useful products, such as larvae converting waste into protein.

6. Circular Economy:
   - Economic system aimed at minimizing waste and maximizing resource use efficiency.

7. Protein Transition:
   - Shift towards alternative protein sources, like insect proteins, due to sustainability concerns.

8. Aquaculture:
   - Farming of aquatic organisms, where maggot meal can be used as a sustainable feed source.

9. Biorefinery:
  - Facility or process that integrates biomass conversion technologies to produce bio-based products.

10. Chitin:
  - Polysaccharide found in the exoskeleton of insects, including BSF, used in bioplastics and medical applications.

11. Insect Farming:
  - Rearing insects for food, feed, or other products, such as BSF farming for protein production.

12. Entomophagy:
  - Human consumption of insects as food, promoted by the nutritional benefits of maggot meal.

13. Vermicomposting:
  - Composting process using worms, contrasting with BSF larvae's ability to digest broader types of waste.

14. Larval Biomass:
  - Collective weight of BSF larvae, containing protein and lipids used in feed and biotechnological applications.

15. Waste Valorization:
  - Conversion of waste materials into valuable products, exemplified by BSF larvae transforming organic waste.

16. Carbon Footprint:
  - Total greenhouse gas emissions caused by an individual, organization, event, or product.

17. Nutrient Cycling:
  - Movement and exchange of organic and inorganic matter back into the production of living matter.

18. Organic Waste:
  - Biodegradable waste materials derived from living organisms, suitable for BSF larvae feeding.

19. Biogas Production:
   - Production of methane-rich gas through anaerobic digestion of organic materials, potentially using BSF waste.

20. Insect Protein:
   - Protein derived from insects, increasingly recognized for its sustainability and nutritional benefits.

21. Sustainability:
   - Practice of meeting current needs without compromising future generations' ability to meet their needs.

22. Bioeconomy:
   - Economic system based on sustainable biomass production and conversion into value-added products.

23. Biopesticide:
   - Pest control agent derived from natural materials, potentially from BSF larvae extracts.

24. Food Security:
   - Condition ensuring all people have physical, social, and economic access to sufficient, safe, and nutritious food.

25. Integrated Pest Management (IPM):
   - Strategy using multiple pest control methods, potentially including BSF larvae for biocontrol.

26. Larval Development:
   - Stages from egg to adult in insect life cycles, manipulated for efficient BSF farming.

27. Feedstock:
   - Raw material used for feeding BSF larvae, typically organic wastes like food scraps or manure.

28. Insect Frass:
   - Excrement of insects, utilized as fertilizer due to its nutrient content.

29. Entomotoxicology:
   - Study of toxins in insects, relevant to BSF larvae's potential role in waste detoxification.

30. Bioprospecting:
   - Exploration of biodiversity for commercially valuable genetic resources, applicable to BSF for bioactive compounds.

31. Myiasis:
   - Infestation of living vertebrates with dipterous larvae, such as BSF, managed for beneficial purposes.

32. Biocontrol:
   - Use of natural enemies, including BSF larvae, for controlling pests in agriculture.

33. Insect-Derived Biopolymers:
   - Polymers derived from insects, such as chitin and proteins from BS

REFERENCE

(1) Maggot Farming – The New Million Naira Breakthrough Business. https://bizwatchnigeria.ng/maggot-farming-the-new-million-naira-breakthrough-business/.

(2) Sustainable Commercial Maggot Production (Maggotry) For Animal .... https://www.academia.edu/6984948/Sustainable_Commercial_Maggot_Production_Maggotry_For_Animal_and_Aquafeeds_In_Rivers_State_South_South_Nigeria.

(3) All You Need To Know About Maggot Farming - BizWatchNigeria.Ng. https://bizwatchnigeria.ng/all-you-need-to-know-about-maggot-farming/.

(4) Maggot farming - Wikipedia. https://en.wikipedia.org/wiki/Maggot_farming.

(5)"Sex Ratio of Black Soldier Fly (Hermetia illucens) Eggs" by M. A. Alam et al., 2020

(6) "Influence of Diet on Sex Ratio of Black Soldier Fly (Hermetia illucens)" by A. M. M. Islam et al., 2019

(7)"Sex Ratio and Reproductive Biology of Black Soldier Fly (Hermetia illucens)" by S. K. Singh et al., 2018

8. Waldron, K. W. (Ed.). (2019). *Handbook of waste management and co-product recovery in food processing: Volume 2.* Woodhead Publishing.

9. Gullan, P. J., & Cranston, P. S. (2014). *The insects: An outline of entomology (5th ed.).* Wiley-Blackwell.

10. Price, P. W., Denno, R. F., Eubanks, M. D., & Finke, D. L. (2011). *Insect ecology: Behavior, populations and communities.* Cambridge University Press.

11. Bondari K. Sheppard D. C. 1987. *Soldier fly, Hermetia illucens L., larvae as feed for channel catfish, Ictalurus puctatus Rafinesque, and blue tilapia, Oreochromis aureus (Steindachner). Aquaculture and Fisheries Mgt.*18: 209–20.Google Scholar

12. Copello A. 1926. *Biologia de Hermetia illucens* Latr. Rev. Sco. Entomol. Argent.1: 23–27.Google Scholar

13. Sheppard D. C. Newton G. L. . 2000. *Valuable by-products of a manure management system using the black soldier fly - a literature review with some current results.* Proceedings, 8th International Symposium - Animal, Agricultural and Food Processing Wastes, 9–11 Octoer 2000. Des Moines, IA. American Society of Agricultural Engineering, St. Joseph, MI.

14. BioSystems. (2021). Black soldier fly production manual: A guide for black soldier fly production. BioSystems.

15. Dossey, A. T., Morales-Ramos, J. A., & Rojas, M. G. (Eds.). (2016). *Insects as sustainable food ingredients: Production, processing and food applications.* Academic Press.

16. Halloran, A., Flore, R., Vantomme, P., & Roos, N. (Eds.). (2018). *Edible insects in sustainable food systems.* Springer Nature.

17. Huis, A. van, Tomberlin, J. K., & Lauck, G. J. M. (Eds.). (2018). *Insects as food and feed: From production to consumption.* Wageningen Academic Publishers.

18. Nation, J. L. (2008). *Insect physiology and biochemistry* (3rd ed.). CRC Press.

19. Rojas, M. G., & Morales-Ramos, J. A. (Eds.). (2014). *Insects as sustainable resources for bioproducts, biomaterials, and bioenergy.* Elsevier.

20. Schowalter, T. D. (Ed.). (2006). *Insects and sustainability of ecosystem services.* Springer.

AUTHOR'S BIO

Oko Obasi is a highly sought after leadership, motivation, stress management, team building expert, success coach, professional speaker, and certified Life Coach.

A Public Policy and Legislative affairs consultant, Oko Obasi is the founder and managing partner of Didarivian Inc., a business consulting and advisory services firm based in Abuja, Nigeria.

Presently he works at the National Assembly, Abuja as a Senior Legislative Aide. He has been instrumental to drafting many legislations that have become Acts of Parliament in Nigeria.

As a seasoned public policy consultant and Senior Legislative Aide at the National Assembly, Abuja, Oko works directly with Top government officials such as Senators, Ministers, Diplomats and Captains of Industries on the issues and challenges that they face in each stage of their official work life cycle.

A politician, human activist, a natural born leader, mentor, and inspirational motivator. A devote Christian and a family man .He is happily married to Ifeyinwa Peace Igbokwe

Obasi. They couple are blessed with five handsome sons.

Oko is author of several books including: Marriage Reloaded: From Eden to Zion, Effective Communication in Hospital Environment, Goshen Exclusion, Pinochio Syndrome, Eunique: A guide to understanding Unique Selling Points (USP), Take Action: Shoot from plans to Execution, Sleep like a Baby, wake like Royalty, The Intelligent Hospital, Thrive Through Tough Times, The Healthy Place, Beyond Pearls: Value of Women, and Just one day(a novel). All these wonderful books are available at Amazon.com.

URGENT PLEA!

Thank You For Reading My Book!

I really appreciate all of your feedback and

I love hearing what you have to say.

I need your input to make the next version of this

book and my future books better.

Please take two minutes now to
leave a helpful review on
Amazon letting me know what you
thought of the book:

https://www.amazon.com/author/okoobasi

Thanks so much!

- Oko Obasi

www.ingramcontent.com/pod-product-compliance
Lightning Source LLC
Chambersburg PA
CBHW071927210526
45479CB00002B/580